# AGAINST THE CURRENT

*Centre de Sciences Humaines* (Centre for Social Sciences and Humanities): Created in 1989 and based in New Delhi, the CSH is part of the network of research centres of the French Ministry of Foreign Affairs. The Centre's research work is primarily oriented towards the study of issues concerning the contemporary dynamics of development in India and South Asia. The activities of the Centre are focused on four main themes, namely: Economic growth and sustainable development, International and regional relations, Institutional structures and political constructions of identity and Urban dynamics.

Centre de Sciences Humaines, 2, Aurangzeb Road, New Delhi 110011,
India, Tel: (91) 11 3041 00 70, Fax: (91 11) 3041 00 79
E-mail: infos@csh-delhi.com
Website: http://www.csh-delhi.com

*Institut Français de Pondichéry* (French Institute of Pondicherry): Created in 1955, the IFP is a multidisciplinary research and advanced educational institute. Major research works are focusing on Sanskrit and Tamil languages and literatures—in close collaboration with the Ecole Française d'Extrême-Orient—ecosystems, biodiversity and sustainability, dynamics of population and socio-economic development.

Institut Français de Pondichéry, 11, Saint Louis Street, PB 33
Pondicherry 605001, Tel: (91) 413 2334 168
Fax: (91) 413 2339 534
E-mail: ifpdir@ifpindia.org

# AGAINST THE CURRENT

## Volume III

## Electricity Act and Technical Choices for the Power Sector in India

*Edited by*

PREM K. KALRA and JOËL RUET

MANOHAR

CENTRE DE SCIENCES HUMAINES

2006

First published 2006



ISBN 81-7304-684-0

*Published by*
Ajay Kumar Jain *for*
Manohar Publishers & Distributors
4753/23 Ansari Road, Daryaganj
New Delhi 110 002

*Typeset at*
Digigrafics
New Delhi 110 049

*Printed at*
Lordson Publishers Pvt. Ltd.
Delhi 110 007

*Distributed in South Asia by*
FOUNDATION BOOKS
4381/4, Ansari Road
Daryaganj, New Delhi 110 002
and its branches at Mumbai, Hyderabad,
Bangalore, Chennai, Kolkata

# Contents

# Acknowledgements

The volume is the third in a series on power reforms in India. The series attempts at understanding (*i*) the organizational tasks, (*ii*) the tariffs aspects, (*iii*) the role of the private sector, (*iv*) the role of technology in the complex, variegated, state-specific, Indian scenario. And formulating clear directions along these four lines will imply understanding their interactions, their meshing together.

This book focuses on the fourth point with the backdrop of the first two volumes of the series. It includes studies and papers presented and discussed at a seminar in Kanpur in September 2003 that was jointly organized by the Indian Institute of Technology of Kanpur, the Centre de Sciences Humaines of New Delhi, Infrastructure Development Finance Company, Institute of Electrical and Electronics Engineers (UP Section), and sponsored by Deloitte Touch Tohmatsu, Power Finance Corporation Ltd. and National Thermal Power Corporation. The editors wish to thank the participants and authors for their contributions and comments. Our warm thanks also go to Attreyee Roy Chowdhury, Publications In-charge at the CSH, for her speedy and cheerful help in this publication.

*December 2005* PREM K. KALRA, Kanpur — JOËL RUET, London

# Acknowledgements

[illegible]

# Contributors

N. Balasubramanian is a PhD student at Indian Institute of Management, Ahmedabad.

R.K. Belapurkar is Executive Director, Bharat Heavy Electricals Limited.

Yogesh Bichpuriya is a M.Tech Student, at Dayalbagh Educational Institute Faculty of Engineering, Agra.

Debashish Biswas is Consultant with Price Waterhouse Coopers, Calcutta (PWC).

Parul Goyal is a PhD student at the Indian Institute of Technology, Delhi.

Sanjay Gupta is a consultant with Infosys Ltd, Bangalore.

Prem K. Kalra is Professor at the Indian Institute of Technology, Kanpur.

D.P. Kothari is Professor and Deputy Director at the Indian Institute of Technology, Delhi.

Mahendra Kumar is Director (Business Development), Power Trading Corporation of India.

Sudha Mahalingam is a Researcher at Institute for Defence and Strategic Analysis.

Deepa Menon-Choudhary is a CSH & Indian Council of Social Science Research (ICSSR) Research Fellow, Centre de Sciences Humaines, New Delhi.

Joël Ruet is Marie Curie Research Fellow at the London School of Economics and Research Associate, CERNA, Ecole des Mines, Paris.

Tirthankar Nag is a Consultant at Credit Rating Information Services India Limited (CRISIL), Mumbai.

A.K. Saxena is Professor, Dayalbagh Educational Institute, Faculty of Engineering, Agra.

P.R. Shukla is Professor, Indian Institute of Management, Public Systems Group, Ahmedabad.

V.P. Singh is at Hindustan Aeronautics Limited (HAL), Nasik.

G. Subba Rao is Chairman of the Gujarat Electricity Regulatory Commission.

# *Introduction*

PREM K. KALRA AND JOËL RUET

India's history of economic regulation entered a new era when, in the power sector, the model of 'independent regulation' for utilities got enacted through the Electricity Regulatory Commissions Act, 1998, then followed by the new Electricty Act 2003. These Acts were instrumental in making India engage with a new political economy of reporting, tariff setting, economic and technical regulations, moving away from the old system of administered economy, by bringing in another set of players: the civil society, and soon the private actors. However, in parallel to this with patient, decade-long, efforts to frame and implement these two Acts, a series of other reforms went ahead: a few property reforms (privatization), structural reforms (some forms of unbundling that were to be noticed before 2003), and many 'project-based' reforms (like the Accelerated Power Distribution Reform Programme). Even though several reformers would obviously see the continuum and even more the intrinsic unity of these series of reforms—the World Bank even developing a particular 'package' of these for the States of Andhra Pradesh, Haryana, Orissa—most would long to see or wish to see these reforms in their isolation.

Typically, utilities would focus on the project-based or project-related aspects of reforms. Signing for larger reforms in the long run was, as often, a desperate move to keep accessing means for new engineer-oriented projects in the short run. For quite some time, the slow implementation of a regulatory process would be for most of the utilities at best an incidental feature of a new way to bargain for resources with the bureaucracy, at worst a distant impediment that would always be sorted out with the help of the bureaucracy. However, after a period of five years of this observationist regime (during which the regulators themselves forged their own regulatory capacities and learning), the 2003 Act got passed. Along with it, was the compulsion to provide for rules and regulations to facilitate open access to the network for external players. Along with this, the real impact of the regulatory process were soon to be felt. Especially, the regulatory need for tariff fixation would lead to a better knowledge on costs, that would in turn necessitate a real reporting on operational factors and on cost assessment of techniques and technology.

An economic revolution had entered the core technical functioning of what had been the most engineer-driven network industry in India, the power sector. Indeed, some of the key provisions of the Act are that 'Generation companies, distribution licensees and trading licensees can demand Open Access to transmission facilities right from the time of notification of the law on June 10, 2003', and that 'Owners of Captive Generating Plants who use the grid for transmitting energy from their generating plant to their point of use have the right to Open Access from the date of notification' (UPERC, 2003). As the Uttar Pradesh Electricity Regulatory Commission (UPERC), one of the pioneering commissions of the country, then put it right away, the direct implications were thus that, for implementation 'as is evident (...), as a part of the mechanism, the Commission would have to specify the applicability and determine the principles for computing the surcharges over and above the wheeling tariff. It is also the duty of the Commission to facilitate Open Access and adjudicate on the adequacy of transmission facilities in event of dispute', and that 'Transmission capacity constraint is a major problem on account of chronic under-investment in the past. There are metering and other infrastructure constraints to be addressed. Further Open Access is only a facilitator for trading of power. Power trading itself will require revamp of scheduling and dispatch arrangements and implementation of balancing and settlement mechanisms'. The regulatory commissions got to have their say on virtually all technical matters within the utilities. The biggest *chasse gardée* of the State Electricity Boards engineers had not resisted.

The series *Against the Current* has detailed already quite a few of the aspects related to reforms of the power sector in India, and this current volume focuses on the relevance of the Act in framing the new Indian power system, with a special assessment of current margins in technological enhancement of the sector. The pros and cons of the Act are discussed here, both in terms of conception and implementation. The crux of the matter is that the debate on the Act brought the issue of the regulation of the sector, and inter-alia, of even the technical structure and nature of SEBs into the forum of public debate.

After several solutions having been attempted on their own and nearly in 'isolation' to each other, the public debate on reforms of the power sector in India is today forced to recognize the intricacies of the problem, and to look at the externalities of each kind of reform, namely: managerial and organizational, technological, financial and on tariffs. Organizational restructuring enables implementing managerial procedures that otherwise may not be possible, or allows considering some technical choices that otherwise would remain hidden and un-accessed in a centralized planning system. These latter elements in turn may provide financial ease in the long run that might call for substantially lesser constraints on tariffs. In reverse, better structured tariffs may lead to smoothening the revenues and thus to ease the day-to-day management. These are few examples, that we consistently develop in our series, the general idea being that, having inter alia recognized (see Ruet, 2005) the need for redefining the meaning, the nature and the socio-economic deliverables of a utility-type industry, this is the right time to recognize that the problem of looking at revenues matching costs is multi-dimensional in nature.

More precisely, the first volume of our series insisted upon the mechanisms through which managerial choices and procedures, have an impact on both costs and revenues, and reforms on these lines provide avenues for re-thinking the cost structures, the investment needs, the revenue pattern and touch upon both an improved use value and social value for electricity (*Against the Current: Organizational Restructuring of State Electricity Boards*). The second volume dealt with tariffs and the effective role of the private sector, linking up tariffs and finance, as well as delinking reforms from the privatization 'model' as attempted in Delhi (*Against the Current: Fixing Tariffs, Finance and Competition for the Power Sector in India*). Here, in the framework of the new Act, we analyse how technical choices and technical upgradation plays a similar role.

## REGULATION AND TECHNOLOGY

Regulation and technology interact, regulations are not mere rules of the game, and neither is technology a mere given. Let us look at a few contemporary examples from India.

Pilot Distribution Circles have been identified by the Ministry of Power, for which the 'Accelerated Power Development Programme' has been initiated in the financial year 2000-1. This provides that a 'focused investment programme has been initiated in 63 identified distribution circles', the Ministry of Power (2002) reports. These projects still need to be evaluated, but they provide help in the right way, and should be specially strengthened for those States which was't to go in for a non-private reform. Now, they are mostly based on technical upgradation of the existing systems. They do provide a short run improvement in the functioning of those systems. The very open question is however, for anybody serious about the power sector, will these sustain 'after the investment'? Definitely, setting indicators, setting operational parameters, setting rules for investment and checks, either directly through control and command or indirectly through the right economic incentives is, in the future, something for which the regulatory commissions need to build a capacity.

Studies like Alagh, Shah & Shah (1998) had shown that if unplanned shutdowns at the generation unit level could be reduced, substantial savings would emerge. Examining the record of each generator, they found that the major hours of shutdown were on account of tube failures. If the concerned Board could improve the maintenance of tubes, the cost of production of electricity would go down on account of three reasons—reduction in energy input per unit of output, overhead costs distributed over larger volumes of production and larger revenue sales from the same capacity. Alagh et al. estimate this saving at 40 paise per unit if GEB could achieve the maintenance norms set by the NTPC. This example goes with the margins of improvement already discussed in distribution, transformers repair and many others (see *Against the Current I*). The key issue will be the competency-building in SERCs and the availability of resources for them to carry counter studies to push for change of procedures. That way, the recently issued tariff orders show a progress in the SERCs mastering technical and financial data provided by the Boards.

Another large scope for technical improvements in capacity utilization and cost reduction, resides in the creation of a market for reactive capacity. Outsourcing of auxiliary services has been explored in many countries but not in India. It is however not out of reach for any Regulatory Commission to regulate the provision of these.

Last example, in Uttar Pradesh, the Commission has been able to differentiate short run and long run margins in losses reduction, by taking into account the ratio between Low Voltage lines and High Voltage lines, or to mathematically model the minimization of network costs (see UPERC, 2003).

Technologies can serve regulatory aims, while regulation can help technological choices to emerge.

## PLAN OF THE BOOK

The book is organized in three parts: 'Electricity Act and Efficiency in Generation, Distribution', that generally comments the Act, 'Technical Choices and Management', that develops ideas and ways to improve technological management of the sector, and 'Macro-Level Aspects: What to Do as Reforms?' that takes perspective of these to conclude on the general impact of reforms as well as energy and environmental choices to be made.

In Chapter 1, called 'Electricity Act and 'Enterprisation' of SEBs', Joël Ruet recalls the main implications but also gaps provided by the Act regarding the possibility to implement innovative structures and technologies in SEBs. In Chapter 2, 'Central Act and State Acts', G. Subba Rao gives a practioner's view of the way ahead and the transition phase for implementing the Act. In Chapter 3, 'Benchmarking of the Electricity Distribution Companies in India', Prem K. Kalra, Vipin P. Singh and Yogesh K. Bichpurya set the path for implementing a regulation aiming at benchmarking SEBs, and in Chapter 4, 'Markets Through the Back Door', Sudha Mahalingam concludes Part I of the book with the many questions that have arisen after the passing of the Act.

Part II starts with Chapter 5, 'Need for Distribution System Reforms and Customer Training', where A.K. Saxena and P.K. Kalra provide arenas for a better integration of demand and supply. In Chapter 6, 'Application of Information Technology to Improve Performance in Power Distribution', Sanjay Gupta shows how the new economy can serve the old one, and suggests how tools go along with internal structures. In Chapter 7, 'Broad Technology Choices and Technology Mix for the Power Sector', R.K. Belapurkar, adopts a far-ranging perspective on comparing and putting together the large generation options, introducing the debate on energy sources through a technological point of view, implicitly leaving socio-economic impacts for the political debate but actually providing technical elements for it. In Chapter 8, 'Voltage Control Through Reactive Power Management', Mahendra Kumar exemplifies the quality aspects of the electricity problem in India, while in both Chapters 9, 'Transmission Line Congestion Management Using Bid-Areas Division Techniques', by D.P. Kothari and Parul Goyal, and Chapter 10, 'Integrated Resource Planning in Supply Side Management in Power System', by Prem K. Kalra, Yogesh K. Bichpuriya and Vipin P. Singh, we take a more systemic approach to the network seen as the central point of the whole electric system.

Part III takes account of the several contributions in the volume to set them in the larger perspective. In Chapter 11, N. Balasubramanian shows 'A Benchmarking of Performance of Indian State-Level Electricity Utilities in Post-Liberalization Period'. In Chapter 12, 'Generation Choices: A Retrospective Account of the Ninth Plan', Joël Ruet concludes the same through a more direct techno-economic analysis, while in Chapter 13, 'Electricity Reforms, Firm-level Responses and Environmental Implications', Deepa Menon-Choudhary, P.R. Shukla, Tirthankar Nag and Debashish Biswas show the environmental dimension, still unadressed really in India, of the power sector. In Chapter 14, Ruet brings together the various conclusions presented by the studies of this book, in order to provide comprehensive outline for furthering of reforms.

## REFERENCES

Alagh, Y.K., Jayanti Shah and V.K. Shah, *Power Economics in Gujarat*, New Delhi: Har-Anand, 1998.

Ruet, Joël, *Privatizing Private Cuts? Ownership and Reform of State Electricity Boards* in India, New Delhi: Academic Foundation, 2005.

UPERC, 'Discussion Paper on Issues in Open Access Implementation', consultation paper, Lucknow: Uttar Pradesh Electricity Regulatory Commission, September 2003.

# PART I

# ELECTRICITY ACT AND EFFICIENCY IN GENERATION, DISTRIBUTION

# 1

# *Electricity Act and 'Enterprisation' of the State Electricity Boards*

Joël Ruet

If continuing for years, any crisis situation may develop some adaptative structural aspects as well as lasting self-reenforcing mechanisms that can sometimes be the source of a vicious circle. In this respect, and beyond technological choices, economic constraints, tariff aspects, dealt elsewhere in this collection of essays, let us ask two questions about State Electricity Boards (SEB) in India:

(i) What are the structural factors and type of procedures, forms of budgetary and hierarchical power structures, etc., in the organization and the decision-making processes within SEBs either leading to, or even reinforcing the problems faced by the power sector?
(ii) What could be the methods to reverse the internal rationality of SEBs in order to implement a reform which would have some chances of success, and what is the impact of current reforms in this respect?

These are questions that need to be addressed in their generality. This chapter, in a limited space, recalls some information-based decision-making procedural bottlenecks in SEBs reaching economic efficiency. It analyses these in the light of the implications of the Electricity Act 2003. Indeed, a new Alectricity Act (2003) has been enacted which replaces the Electricity Act of 1948 (as well as the Act from 1910) and the Regulatory Commission Act (1998), extending the provisions of the latter, and breaking away from its SEB-based organization promote a competition-based sector in the future.

To serve as a framework linked to the first question, we call 'enterprisation' a process that would enable creating and using decentralized information to reach economic efficiency (Section 1). To answer the second question, we explore how and with what limitations the Electricity Act aids in solving these bottlenecks, thus 'enterprising' the SEBs (Section 2).

## INTERNAL ORGANIZATION: SEBs AS ADMINISTRATIONS, NATURE AND CONSEQUENCES ON REFORM

Beyond 'political interference' and tariff structure, the organizational characteristics of SEBs (over-centralization, technical focus instead of cost-based decision-making, to name a few) explains to a large extent their inefficiency (for an extensive survey of SEBs across the country and an articulate analysis of these, see Ruet: 2001, 2002 and 2005). A non-integrated (between physical and accounting elements) information system, the absence of internal allocation of discretionary power (budgetary and hierarchical), and the pre-eminence of paper work (what is decided on paper counts more than what actually happens) prevent the operational units to come out of the framework of procedures in decision-making. Let us detail these.

All decision-making in SEBs revolves around public accountancy, where actual clearing is done at the headquarters level, *ex-post*. That way, the budget of each unit is not strictly 'allocated' to it in a granted manner. The so-called budget allocation does not even allow the field the use of discretionary power to arbitrate within two similar actions in slightly different places or at slightly different times. A decision is separate from any other one, and each follows a separate path. In enterprises, conversely, the budget is classically designed to be a managerial tool for the headquarters, allowing a monitoring of the totality of expenses while decentralizing the budgetary constraint. This decentralization can be possible only if the credits in the budget serve the purpose of being—at least partially—monetary equivalents between the actions one needs to arbitrate.

SEBs function with an administrative-type budget, a so-called 'allocation'. In a real enterprise, as opposed to SEBs, decision-making-cum-accounting units are defined. Once granted a budget, the finances are actually transferred, at fixed times, to the local accounts. Then these units have rights and responsibilities over decisions and related expenses up to a fixed level of delegated financial power. They later on report on these expenses. This allows fast and decentralized decisions, and puts the burden of responsibility (and hence acts as an incentive) on local managers. For instance, an enterprise like the French public power company EDF has since long applied this system. This does not exist in SEBs. Even operational day-to-day decisions cannot be taken efficiently and rapidly at the local level. In the constrained framework of public accountancy, the goal of managerial freedom is not met. This is a book-currency budget, or a paper budget; it is not actually allocated. The paper budget at stake in SEBs simply precludes this kind of decentralization.

To sum up, in the framework of the book-currency budget, the power of decisions left to the field is very low. Further, the power of making decisions that imply direct purchase from the market and not through the internal store system, is even more limited, next to non-existent. It can be concluded from this that:

(i) the actual discretion on external purchase is very limited;
(ii) internal allocation of inputs is subject to stringent conditions; and
(iii) in both cases, the number of clearances at various levels is high; prior to a decision, at least three levels are involved in preparing a technical plus administrative sanction. Following that, it still has to be cleared by headquarters under the budgetary process or the authorization of expenditure through the bank.

This leads to the fact that at every level of the hierarchy, officers spend a substantial part of their time clearing formal requests. This need for multiple sanction implies that every officer has tall piles of documents to sign, in which a clerk shows the officer where to sign. The fact that this system can daily be seen in the offices, and that basic errors remain, shows that these reports are not read when signed. Each day is busy with tides of helpers moving piles of files. In addition, the time taken to have a decision cleared can be very long. On close examination, it is clear that the separation of the two levels of sanction and the framing of the decision process within a book currency system leads to the *de facto* absence of formal discretionary powers at the field level.

The actual activity of SEBs is more technical. Ensuring physical continuity of supply is the most time consuming activity. This ranges from ensuring that generation plants will deliver energy, down to reacting in case of power cuts in distribution. Within organizational characteristics of SEBs, the stress on continuity of supply justifies (*i*) executive instructions from the governmental; (*ii*) a relative disinterest in other activities on the part of headquarters; (*iii*) quasi-exclusive priority for day-to-day aspects rather than long-term aspects; and (*iv*) even misreporting in some circumstances.

The key aspect is that, given the systemic unavailability of discretionary power or information, relying on paper work instead of costs becomes rational. The point here should not be misunderstood. If paper work leads to cost inefficiency, as compared to a situation where costs would be targeted, it does not mean that inefficiency is targeted *per se*, but that strong inertia to reform is in-built in the SEBs' internal organization.

This forecloses the possibility of otherwise profitable decisions, and the decisions follow *de facto* a procedural rationality (defined by Simon, 1976; March 1978): for instance, in a context where economic pertinacity of a decision cannot be proven because of lack of a minimally integrated information system, procedural rationality could mean relying on administrative and technical rules or rules of thumb, rather than looking for innovative, cost-effective measures. In SEBs, the incentive to do so is reinforced by the fact that, at any moment and at any level, executive instructions from the administration can occur. In this fluid framework, the difficulty for managers to anticipate even otherwise predictable events makes legitimate not to consider substantive rationality. Conformation to administrative rules gives higher stability than trying to anticipate external constraints. These interventions need not necessarily be 'systematic'; their possibility itself leads to their anticipation, which is systematized by the agents. An adapted system of procedures emerges from it, each decision will then find its genesis diluted in various points of this system. In that sense the procedural rationality is 'systemic'.

Consequently, reform, the way we define it, should be understood as the willingness to grant a central position to the concept and usage of cost in the decision-making process at a decentralized level, necessarily going through a change in the SEBs' nature, from an administration to an enterprise. We have named this process 'enterprisation'.

DEFINITION

- Enterprisation of SEBs amounts to their proper capitalization through building of assets, that is, the accumulation of capital through their production.
- As a corollary, their reorganization around cost as a central concept is a prerequisite.

A practical corollary to this is the suppression of the existing administrative procedural rationality and hence of the present procedures. Beyond 'political interferences', the real aspects to be suppressed are the structural organizational features of the Indian administration.

Similarly, at the sectoral level, all public undertakings have long followed similar administered paths. The question of sectoral coordination is the following: when the power sector is de-integrated to look for gain in efficiency through direct competition (in generation) or indirect competition (in distribution at the time of issuing operating licenses), there arise technical, commercial, financial and coordination problems. This coordination may continuously address issues ranging from very short-term—questions of security at millisecond levels—to very long-term—planning of a few decades (Joskow and Schmalensee, 1983; Joskow, 1996). The Electricity Act 2003 provides a legal framework for achieving so in the future. Setting up of adapted modes of governance and institutional mechanisms is in itself a subject of study (Glachant, 1997). Introducing the Electricity Act, especially its measures to favour open access, call for redesigning these. The question becomes (*i*) how is this possible with the existing nature and functioning of SEBs, and (*ii*) how SEBs' functioning, in turn, gets altered by the very compulsion to adapt to the Act. In short, does the Act help enterprising the functioning of SEBs. The short-term coordination continues to be ensured according to technical regulations, unaffected by the organizational requirements of the so-called structural reforms. Above all, the question of coordination refers to the issue of long-term investments, of risks and guarantees, faced with stranded costs and specific assets.

In that context, the general debate on the emergence of competition, or the opening up to competition, should rather concern organizations that are already sensitized and which respond to monetary constraints and incentives. Again, trading of energy, organized by the 2003 Act, has concerned less than 1 per cent of the demand in 2002-3 and 4 per cent in 2004-5. It leaves the bulk of SEBs unaffected. In classical terms, competition concerns organizations that already have a minimum sensitivity to cost-efficiency.

The general question becomes: (*i*) how to enterprise SEBs, and (*ii*) who can do this. It is still a very large question. Let us analyse the role of the Electricity Act in this context.

## THE CONCEPT OF ENTERPRISATION AND ANALYSIS OF THE DEBATES ON THE ELECTRICITY ACT 2003

The main features of the Act revolve around the:

- delicensing of generation (and the end of restrictions on captive generation); this materializes in the suppression of the need for Central Electricity Authority (CEA) techno-economic clearance;
- granting of open access (traders/sellers of electricity may use the networks of the utilities/network companies by paying a regulated charge); this is to come into effect immediately as far as transmission is concerned, while the State Electricity Regulatory Commissions (SERC) are given one year for proposing modes of gradual open access in distribution;
- corporatization of remaining SEBs within a year, and unbundling since the newly formed state transmission utilities will be prevented from trading electricity;
- granting of multi-year tariffs, in order to reduce the so-called regulatory risks; and
- an empowerment of the inspection powers of the Regulatory Commissions.

The Act is seen by its inceptors as a way to force unbundling and promote competition by authorizing consumers to shift from the incumbent to the generator or trader of their choice, and, true enough, the Act is indeed ambitious. Shahi (2003) mentions that 'the Act empowers the regulatory institutions to oversee that the utilities function efficiently'. He adds that 'determinants of efficiency, like norms for availability-based tariff (ABT) will ensure that those who fail are adversely affected while those who improve gain financially. For example, if the distribution licensees gradually reduce the technical and commercial losses, the power tariff for consumers will fall'.

Two dimensions are hence present in the Act: strengthening the regulator as well as developing competition. On a similar line, Haldea (2003a) mentions that a notable facet of the new Act, 'following universally best practices', is the open access system under which a set of customers can get together as bulk consumers and purchase power either directly from the generating company or through intermediaries such as traders and distribution companies. If one looks closely at the Act, competition is supposed to come, among others, from section 39(1), which details that 'State Transmission Utility shall not engage in the business of trading in electricity'. Section 39(2)(d) specifies that it will provide 'non-discriminatory open access' to distribution licensees, generating companies and consumers. This ensures a legal compulsion to unbundle SEBs (according to the Act, it was to be completed within a year from the enactment which got postponed . . . ).

However, the ambition of the Act, and its ability to implement it in a phased or rapid manner is precisely what is debated in India. The debate actually revolves around two lines: the feasibility of competition in line with today's achievements of the regulators, and the (unavoidable question of the) role, merits and demerits of the private sector. For instance, Godbole (2003), while agreeing that 'the most important element of the reform strategy contained in the Act is of unbundling of SEBs', argues that the Act 'raises a number of new issues which are likely to pose serious problems in the coming years'. He specifically argues that, given Sections 172 (that provides a possibility that state governments, with the sanction of the Government of India, continue the functions of the SEBs as transmission utilities), and 131 (that gives margins for vesting the property of SEBs in the state government), it is clear that 'no final date has been set for the abolition of SEBs and this decision has been left to the State Governments'. Godbole indeed argues that, politically, the Centre will unlikely terminate a SEB if the concerned State does not wish so. Equally, Subba Rao (2003) argues that 'in a way, the Electricity Act 2003 being a forward looking legislation, may have run quite ahead of the available infrastructure and system capabilities. Proposed innovations like open access and power trading will require many technical preconditions. There has to be enough capacity in transmission lines as well as inter-regional connectivity. Without these prerequisites for moving energy from surplus to deficit areas, open access may not actually materialize. These considerations will also apply to open access in distribution systems.' He adds that 'it may take a while to design and create technical preconditions on ground such as needed hardware, capacities, systems, codes, energy accounting methods and computerized monitoring devices for operating the provisions'. Ultimately, he states that the regulators 'are still fledgling organizations'. Others, like Kapil Sibal, Member of Parliament and former member of the Standing Committee on Energy, even fear that the Electricity Act 2003 will perpetuate private monopoly in the power sector and discourage competition.

There is a risk in entering the realm of the ideological debate, for the Act lacks a history. It is true that the Act does open considerable scope for competition, which will be much welcomed in the future. The current debate in India revolves more around manner how to introduce competition in a phased manner, many arguing that it is prematurate if to be generalized right from now. Such arguments are difficult to clinch straightforward, but, working on the fact that there is a general consensus that the long-term issue is competition, we argue that the framework set in place in this book sheds light on several issues linked with today's phasing of reforms. Especially, one needs to discuss (i) the generalization of the structural reforms in a context where (ii) competition, that is to say, incentives, may be activated, where (iii) the role of the regulator is enhanced, but also where (iv) the bankruptcy risk may be increased and articulated in a specific manner with the debate on privatization. We will discuss the three former points with respect to enterprisation in this chapter, but leave the discussion of the last point (though not the least important) due to lack of space (for this, refer to Ruet, 2004).

## THE *SINE QUA NON*: COMPETITION NEEDS REGULATION, BASED ON INFORMATION, THROUGH THE PREREQUISITE OF ENTERPRISATION

Generally, the setting up of competition is envisaged through establishing open access. However, competition should be disentangled between two levels: generation and distribution. Haldea (2003b) details that 'once you have open access, consumers in south Delhi, say, can decide if they want to buy power from BSES that has the monopoly for the region right now, or from some other company, say Y. Naturally, competition between BSES and Y will lower prices, the same way competition between BSES and Tata Power has driven down tariffs for high-tension consumers in Mumbai'. This statement articulates differently when one considers generation and distribution.

In generation, there may be competition if bulk consumers can choose between different providers of energy. But, as Morris (2003) explains, as of today major economic distortions continue in terms of costing of energy. Captive plants, new plants, old plants, unallocated capacity, all face different rules of pricing, still to be either harmonized or liberalized, but in a converging way, with similar rules for all generators. Morris also underlines more cynically that, in the long run, the risk is that opening up may be necessary to bring the SEBs to collapse, and then to reform. However, if SEBs have to be turned bankrupt, this poses several problems. What would happen if SEBs massively collapse?[1] This scenario is not unlikely, when the Act authorizes most of the customers (hence those who cross-subsidize others) to desert for other generators, like for instance Central PSUs. In a context where SEBs would remain with the subsidized consumers mostly, how can one dream that the size of the private will be enough to cover such a collapse, several authors or practitioners like Godbole (2003) and Sagar (2003), recall? For instance, Godbole specifies that 'it is unlikely that private sector will have the capacity to take over the whole electricity business from SEBs even during the next two decades'. This is why Godbole calls the Act a 'halfway house' and in 'anxiety to (...) be unduly futuristic'. Actually, one might have to remain on a more transitory mode, and in this regard, it is interesting to notice that Haldea (2001) minds the necessary transition by proposing a gradual competitive system where the existing sources 'would not enter the competitive market, and would continue their supplies as per extant agreements. Only new capacity would be subjected to competitive markets to begin with; and competition would thus be introduced at the margins'.

Further, others like Phadke and Rajan (2003) state that 'except for the increased role of the Regulatory Commissions in retail rate setting (...), there are issues of concern in several of the other provisions' of the Act. Their concern is more particularly with the fact that 'in developing countries it would be even more difficult to instil competition and mitigate the exercise of market power. Factors like the small size of markets with large centralized power plants, a tight demand-supply situation with low reserve margins, and virtually no real time demand response mechanisms in place will increase the potential to exercise market power in the generation sector substantially in a developing country market.'

Dealing for instance with the sole 2000 MW coal plant of Chandrapur in Maharashtra, they mention that 'the existence of this large plant is sufficient to make the Maharashtra electricity market highly non-competitive during the peak periods'.

But more structurally, one can wonder on the empowering capacities of 'incentives' and 'pressures' on SEBs to reform themselves. Indeed, and maybe first of all, they *already* are virtually bankrupt! Their. Rate of Return on assets (RoR), which is mandatorily of 3 per cent according to the Electricity Act 1948, has dipped from an already abysmal (–) 19.6 per cent in 1996-7 to a catastrophic (–) 44.1 per cent in 2001-2. True, SEBs are 'under-capitalized' (we know what to think about the concept of capital in their case), but still, at that rhythm, they are just burning their resources. This present situation means, *inter alia*, that SEBs have not been sensitive to 'incentives', which is not a surprise if our thesis is correct that the concept of soft budget constraint is not operative in the analysis of SEBs. In other words, competition is to be seen rather in a framework where, in future, private companies (or public-private partnerships) will be established, and cannot come before enterprisation of SEBs. Let us put it in a different wording. Phadke and Rajan (2003) state that their 'central argument is that the successes of reforms in India will depend critically upon the existence of some sort of restraining or disciplining mechanism in the sector, in the absence of which current efforts will likely result in a transition from inefficient public ownership to profit-gouging monopolies or oligarchies'. They add that, in India, 'in the absence of either effective competition or regulation, whatever benefits privatization has to offer may never reach the common people'. Godbole (2003a) gives an illustrative example: 'In Maharashtra, the private companies Tata Electric Supply Company and the Bombay Suburban Electric Supply have "not filed their annual revenue requirement submissions before the MERC though over three years have elapsed since the setting up of the commission". '

Beyond these measures affecting generation, the Act stipulates that in terms of distribution, several licenses can be granted for the same zone! This distinction between specific distribution (the operation of the local monopoly network) and supply (the competitive commercial use of this network) is, according to what we have described on the level, accuracy and reliability of information, largely premature in India, and this in no way can help enterprising SEBs.

In other words, and as many authors now recognize, the most important developments in the power reforms in India have revolved around the Regulatory Commissions, justly empowered by the new Electricity Act 2003: the much acclaimed competition can come only if credible and empowered Regulatory Commissions have developed capacities to assess the SEBs or their heir entities or the private companies to come. Here lies the credibility of the reform. The Act provides these measures to empower the commissions, and the years to come will tell if and how the Commissions effectively build these capacities. But still, regulation will revolve around its capacities to generate and check data in a dynamic process. As we have detailed, and given the present shape of SEBs, this data building is part of the enterprisation process, and this sheds light on Subba Rao's (2003) call for 'technical pre-conditions'. Here may actually be the wisdom of the Act (along with its manipulability), where section 172 entitling delaying the final unbundling of SEBs. After the 2004 general elections, the new government has of course used this possibility, but one is tempted to believe that the earlier one would have had no other choice but to the the same. . . .

## THE ACT AS A WINDOW FOR EARLY 'MARKET' INITIATIVES: ENTERPRISATION THROUGH LOCAL MARKETS: NO 'ISLANDING', BUT A KICK-OFF EFFECT

The issue of imposing cost-based decisions at the local level can also be addressed through 'helping those who help themselves', following the expression coined by Y.K. Alagh. One way of addressing the problem, is to harness the great vitality of decentralized markets and integrate them with the growth process in the core areas of distribution networks. Organizations and communities which foster limited and well focused areas may be targeted (see Alagh, 2001). Practical experiments have been attempted with some success since 1998. Among them is the much discussed 'Tatkal Scheme', developed initially by the Rajasthan Electricity Board, followed by Gujarat and then Haryana. This consists in giving rapid connections for tubewell consumers having expressed their willingness to pay the full cost. In most of the SEBs, applications for tubewell connections have been pending, the power utilities alleging 'lack of resources'. The potential of this scheme is therefore important: Indeed, Alagh (2001) considers the number of potential applicants in Rajasthan to be at around 18,000 in each district, while there are 60,000 pending applications in Haryana. The Gujarat Electricity Board reports having electrified 29,200 applicants by the end of March 2002 under a Tatkal Scheme, and 61,000 by September 2002. Tatkal connections are all metered, with consumers having agreed to commit to a minimal tariff (Rs. 1.40 per unit in Rajasthan for instance). Similarly, one should promote schemes where a village or industrial estate promise to pay the price of electricity, and then provide the facilities urgently and regularly to them.

In other words, new projects on sub-systems of the power economy should be promoted. This concept of 'islanding' is therefore a constitutive part of an enterprisation strategy, and complements the macro-economic and organizational reform of the entire system. Now that the Electricity Act 2003 is in place, such strategies can easily be implemented within its framework.

On this point, the argument of this book in favour of tying-up with the private sector is as follows. First of all, the various schemes described as part of the enterprisation process, such as allocating financial discretionary power to the field, or building-up an information system that field officers can effectively use, concur in locally complementing the Tatkal Scheme or any islanding measure. Indeed, farmers show a willingness to pay but, for it to sustain, the SEBs have to provide a reliable power. This reliability has to be both in terms of guarantee of supply over time (no power cuts) as well as in quality of supply (voltage). For that, it is necessary that the decisions based on costs and benefits, that is, that the organization is able to buy or produce more energy to fill its commitment towards the Tatkal consumers. Beyond, it can be ensured only through a better information and metering at the local level. Second, the pursuit of quality also implies local power to able at the 11 kV distribution station.

Let us here reframe the way reforms are practically implemented in SEBs. If reforms are seen by some of their advocates as a way to create small entities to be privatized as quickly as possible, then what becomes important is to give some global visibility to supposedly hurried investors. Balance sheet restructuring then becomes a priority. If it is felt as well that substantial improvements cannot disseminate and generalize in the public structure, even though granting and decentralizing of discretionary power, then the reform of the reporting system can be effected with a headquarters-oriented approach only. This is what happened and what can explain that, even though sitting at the field level, superintending engineers have barely met or seen the armies of consultants doing the job, while the Executive Engineers and Sub-Division Officers (SDO) have not met them at all. On the contrary, if the need to enterprise is understood, these reforms will also target the lower levels of the decision-making, and not be designed at the benefit of the sole headquarters. This can in other words explain how the practical implementation of structural reforms has till now taken the form of solving bounded rationality instead of addressing procedural rationality.

The Act is largely a strengthening of the Regulatory Act 1998, but it includes a major element, where (*i*) transfer of 'property or rights' is no longer a taboo, and where 'open access' is to be promoted in a 'phased manner' (Article 42). In that framework, vertical integration is still allowed (implicitly re-integration once privatized, i.e. transferred to a private licensee), but at the condition that unbundled accounts are evolved (Article 51). The accounting normalization of distribution is therefore engaged through this article of the Act. Further, and at the core of our paper, Article 131 is illustrative of the whole part of the Act relative to 'transformation of Boards'. Article 131 specifies that from a Board, a transfer scheme may 'provide for the formation of subsidiaries, joint venture companies, or other schemes of division, amalgamation, merger, reconstruction or arrangements which shall promote the profitability and viability of the resulting entity, ensure economic efficiency, encourage competition, and protect consumer interests'. In practice, the implicit logic of the Act revolves around a long-term horizon of a large private sector, where commission are also to promote competition, and, among all, open access. In open access, any company can wheel its power through the distribution network and sell it to its clients. Paradoxically, this corresponds to a logic where distribution becomes a business of optimizing a network, of reducing its costs, and where integration with a generator is merely a provision allowed by the Act, but not necessarily the norm (for, actually, distribution is a local monopoly and different energy providers can compete through it). This in turn normalizes the function of electricity distribution and totally aligns it with the logic and advantages of the management contracts described earlier.

## CONCLUSION: SHIFT IN PARADIGMS

To conclude, our core argument is that the long neglect of maintenance, quality of service, cost-efficiency and conservation measures has come from an administrative paradigm. We propose an alternative. Shifting from one paradigm to another always requires a catalytic element, and the bridging of political lock-ins. We are confident that the public debate in India is progressing towards a clarification of analysis and an evaluation of past experience, that the technological innovations may serve as a catalyst, and that political lock-ins may be overcome.

As foreseen and analysed by Joskow and Schmalensee (1983), and confirmed since then, the technical change has brought the ability to decentralize technological improvement at the user end. This has broken the long-time paradigmatic vision of the power sector as a naturally vertically integrated sector. The ancient argument was that transmission is a natural monopoly and leads to the integration of the whole sector because of the coordination needs and the many externalities the transmission network has on distribution and generation. Technological change has introduced innovative coordination forms (Finon and Glachant, 2000) and changed the ancient paradigm into a trade-off between the co-ordinative pros and cons of integration (Glachant, 2002). However, the problems of the power sector in formerly administered economies are radically different (Ruet, 2002) and any reform in India has to integrate the issue of 'enterprisation'. Only then, technology can be fully made used of. Enterprisation may fix the targets and the milestones for catching such sources of improvement. At this degree of complexity (including in policy formulation), neither the administrative paradigm nor the 'full-private' paradigm can succeed. As Phadke and Rajan (2003) put it, using at par examples of developed and developing countries and relating these to India, 'ownership change must be looked as one among several options for changing incentives and not as an end goal'. The reduction of the debate to mere ideology at the turn of the 1990s has produced nothing but the worst capacity addition in the recent five-year plans.

In this highly sensitive field, a sound political debate focussing on genuine arguments matters. Political lock-ins are being progressively overcome in India. Regulatory Commissions have been set up through an ordinance, soon changed to an Act. Though they were not compulsory initially, it became so when the 2003 Act was passed. Minimal tariffs for agriculture could not be centrally imposed, neither in 1996 nor in 1998, nor any time later. Today, nearly all the states have adopted their own minimal tariff for agricultural consumption, and a legal minimal national tariff might finally come into being. A transmission Bill opening the field to private investment could be accepted by a Parliamentary Committee, after objections of the left-run states are answered and they are convinced that grid operation would be in public control, even with private investment. Availability tariffs, which had suffered because of judicial cases filed by NTPC, are soon to be back. The case has been rejected by the Supreme Court in November 2002, therefore subjecting all the public generators to the economically sound system of availability tariffs. Finally, the energy efficiency legislation still faces hurdles from the private sector. Working conditions for the private sector to join are still to be found, but this is no longer considered as out of reach. This provides a positive context, but is not sufficient, the book has all along underlined, and the shift in paradigm must take place at the organization level, at the consumer-end level, at the local project level, if it is to sustain. Political power has a crucial role in making these policies start, prove that it sometimes knows how to overcome its own lock-ins. If well-designed and legally enforced at all-levels, enterprisation will then sustain, because it has a sound economic basis, and is no longer limited to a mere administrative measure. But the Act is not the end of the road, and needs being accompanied by development of cost-based information systems at all levels, training and decentralization in SEBs.

## NOTE

1. In a different context, after the collapse of the Soviet system in East Germany, all the 'Combinates' had to be restructured. Hirschhausen mentions that, true, most have been ultimately 'privatized' after restructuring (which saw anyway a large drop in production), but at the net cost of million 900 DM (450 million Euros), when the reformers had initially expected sales for a net positive value of million 2,600 DM. The 3,500 million DM difference had been the inescapable cost of enterprisation.

## REFERENCES

Alagh, Y.K., 'Community & Cooperative Action', keynote address on Development and Governance, in Ginkel and Thakur, *UNU Millenium Series, 2001*, Tokyo: United Nations University, 2001, pp. 73-82.

Finon, Dominique and Jean-Michel Glachant, 'Why do the European Union's Electricity Industry Continue to Differ? A New Institutional Analysis', in Claude Ménard (ed.), *Institution, Contracts, and Organizations: Perspectives from New Institutional Economics*, London: Edward Elgar, 2000, pp. 313-34.

Glachant, Jean-Michel, Le pool d'électricité en Grande-Bretagne: un marché incomplet sous administration privée et publique, Université Paris I, ATOM, 1997.

Glachant, Jean-Michel, 'L'approche néo-institutionelle de la réforme des industries de réseaux', *Revue Economic*, vol. 53, no. 1, Paris, May 2002, pp. 425-36.

Godbole, Madhav, 'Electricity Act, 2003: Questionable Wisdom', *Economic and Political Weekly*, vol. XXXVIII, no. 39, Mumbai, 2003, pp. 4104-10.

Godbole, Madhav, 'Power Sector Woes: No Easy Answer', *Economic and Political Weekly*, Mumbai, September 2003a, pp. 3781-4.

Haldea, Gajendra, 'Whither Electricity Reforms'?, *Economic and Political Weekly*, reprint in Ruet (ed.), *Against the Current: Organizational Restructuring of State Electricity Boards*, New Delhi: Manohar, 2003.

Haldea, Gajendra, Interview in the *Business Line*, 20 June 2003a.

———, interview in the *Business Standard*, 16 October 2003b.

Joskow, Paul and Richard Schmalensee, *Markets for Power, An Analysis of Electrical Utility Deregulation*, Boston: MIT Press, 1983.

Joskow, Paul, 'Introducing Competition into Regulated Network Industries: From Hierachies to Markets in Electricity', *Journal of Corporate change*, vol. 5, no. 2, 1996.

March, James, 'Bounded Rationality, Ambiguity, and the Engineering of Choice', *Bell Journal of Economics*, 9, 1978, pp. 587-608.

Morris, Sebastian, 'The Electricity Act 2003, Implications for the Development of the Sector', presentation made at the seminar Power Sector Reforms: Technological Aspects, co-organized by IIT-Kanpur and CSH-Delhi, September 2003.

Phadke, Amol and Sudhir Chella Rajan, 'Electricity Reforms in India: Not Too Late to Go Back to the Drawing Board', *Economic and Political Weekly*, vol. XXXVIII, no. 39, 2003, pp. 3061-72.

Ruet, Joël, 'Winners and Losers of the Reform of SEBs, An Organizational Study', Occasional paper, New Delhi: CSH, 2001.

———, 'A New Paradigm for the Indian Power Sector', in R.K. Mishra and G. de Bernis (eds.), *Privatisation of Public Enterprises*, New Delhi: Manak, 2002, pp. 220-62.

———, *Privatising Power Cuts*? Ownership and Reform of State *Electricity Boards in India*, New Delhi: Academic Foundation, 2005.

Sagar, Jagdish, 'The Unbundling and Privatization of DVB', paper presented at the Brainstorming Workshop on Power Sector Reforms, Bangalore: IIM, 11-12 April 2003.

Shahi, R.V., Interview, http://www.renewingindia.org/news1_june3_elecact.html, 2003.

Simon, Herbert, 'From Procedural to Substantive Rationality', in S. Latsis (ed.), *Methods and Appraisals in Economics*, Cambridge: Cambridge University Press, 1976, pp. 129-48.

Subba Rao, G., 'Issue in Electricity Bill 2003', Keynote Address, IIT-CSH Workshop on Power Reforms: Technological and Financial, 4-5 September 2003, IIT Kanpur.

# 2

# *Central Act and State Acts*

G. SUBBA RAO

The issues central to power sector reforms have been addressed through the Electricity Act 2003 which is undoubtedly a landmark legislation. Besides the Central Act, there are also eight State Acts listed in the Schedule to the Central Act. In Gujarat, the State Assembly passed the Gujarat Electricity Industry (Reorganization and Regulation) Act 2003. We discuss the implementation of the Act, as seen from a practitioner's point of view.

## INTRODUCTION

In Gujarat the operation of the Central Act was postponed by six months till 9 December 2003. As regards the relationship between the Central Act and the State Acts, four situations can emerge. First, the Central Act and the State Acts may contain inconsistent provisions. In this case, the Central Act will override the State Acts. Second, the Central Act may contain provisions which are not found in the State Acts or vice versa. Since such provisions are non-overlapping, it seems reasonable to suppose that there will be no conflict. As a matter of fact, the Central Act, at least in comparison with the Gujarat Act, has many extra provisions. Finally both the Acts may contain provisions about a particular area without there being any explicit conflict. In this situation, the principle of harmonious interpretation of the two acts will have to be followed.

A noteworthy feature about the Electricity Act 2003 is that it requires elaborate subordinate legislation by way of rules and regulations. This is also seen in Gujarat Electricity Industry (Reorganization and Regulation) Act 2003. The Central Act stipulates that the Government of India (GOI) should bring out policy documents like the National Electricity Policy, National Electricity Plan, Tariff Policy, and Policies covering grid-based or off-grid-based rural electrification. In pursuance of these provisions, the Ministry of Power has already placed a discussion paper on 'Tariff Policy' on their website. The Central Electricity Regulatory Commission (CERC) has placed on its website discussion papers on 'Terms and Conditions of (Electricity) Supply' and a concept paper on 'Open Access'. Regulations have to cover a wide area including grid code, performance standards, modalities for operating power systems and for providing open access. I do not want to list out the regulations or get into their details. Suffice it to say that the final picture of the Regulatory Framework will emerge after all these tasks are accomplished. The Gujarat Electricity Regulatory Commission (GERC) has already set in motion the process for engaging a consultant for preparation of regulations.

The new Act is a path-breaking effort for creating enabling routes for power sector reforms. Its main aims can be described as: promoting competition, protecting consumer interests and providing electricity to all, or what may be called a universal obligation. As compared to the state Acts, the Central Act is more comprehensive and radical in pursuing innovative reforms. Besides introducing new reform-oriented provisions, it has amalgamated three existing Acts: the Indian Electricity Act, the Electricity Supply Act and the Electricity Regulatory Commissions Act. It is noteworthy that the Electricity Act has greatly enhanced the regulatory powers of the Commissions. Rather than getting into these details, we may at this stage look at its new features.

## MAIN FEATURES OF THE ACT AND THE CENTRAL ROLE OF REGULATORS

Under the new Act, generation of power has been delicensed and the earlier requirement of techno-economic clearance for generating plants by CEA has been dispensed with. Similarly, restrictions on captive generation have been removed. Open access has been allowed with immediate effect in transmission. The SERCs have to come out with a programme for introducing open access in distribution in a phased manner. Captive units can immediately access the state grid, subject to satisfying the technical conditions of grid connectivity and payment of wheeling charges for moving energy from the captive plant to the factories located elsewhere. By way of contrast, consumers who want to utilize open access will have to pay besides wheeling charges (1) a surcharge covering current level of their contribution to cross-subsidy and (2) an additional surcharge towards fixed cost of existing licensees. The Act introduces provisions for trading in electricity. Another important feature is the provision for grid-based or off-grid-based self-contained rural distribution systems.

The Act brings about important changes in the organization of the electricity industry. As is well known, in many states, electricity industry is still a vertically integrated monopoly which combines the functions of generation, transmission and distribution. While some progress has been made towards unbundling State Electricity Boards (SEB), the move towards privatization has been far slower. Under the Central Act, SEBs have to mandatorily become corporate entities within one year. The new Act prohibits a transmission utility from engaging in trading. These provisions to some extent will automatically lead to the unbundling of SEBs. The SEBs at present not only function as transmission utilities but also purchase power from generating stations for distribution, or in some cases resale to licensees. In view of the above-cited provisions, this position will have to change. Incidentally this will necessitate restructuring of transcos formed in some states as part of the reform process.

The Act lays down that the Central and State transmission utilities should be government companies. As part of the move towards introducing competition, the new Act provides for multiple distribution licensees. Multi-year tariff is another notable feature of the new Act. It tries to separate the transmission function from that of system operation by creating national, regional and State load dispatch centres. These provisions will also help in introducing open access and trading in electricity by removing some of the conflicts of interest inherent in the combination of disparate functions in the same utility.

This paper will now address some of the issues which are likely to arise in putting the provisions of the Act into actual practice. The intention is merely to flag the issues and in any case these are not insurmountable. As a regulator, one is hesitant to make any simple and quick pronouncements on these matters. As mentioned earlier, it is necessary to study the Act in detail for fully grasping its implications. However, a few general observations will not be out of place.

Perhaps for the first time, the law assigns a developmental role to regulators. For instance, it clearly lays down that the SERCs should 'facilitate' development of electricity markets or 'open access'. Evidently, the Act expects the regulators, other things being equal, to move in the direction of reforms, competition, consumer interest and universal obligation. The Act contains words like 'transparency'. But such words have not been defined, nor have their meanings been elaborated in the judicial pronouncements. But it is clear that the regulators along with other players have to work in such a way as to realize the objectives of the Act. Indeed, in the discussion about the Act, issues like regulatory capabilities, regulatory predictability and regulatory uncertainty have cropped up. A few observations can be made on this aspect including that of the likely general regulatory stance.

The SERCs are deemed to be Civil Courts and their proceedings are judicial proceedings. As already noted, the regulatory proceedings have to be open and transparent. Evidently, while performing their functions, the SERCs have to operate within this legal framework. In fact, this is what will lend credibility and predictability to the proceedings of the Regulatory Commissions. In this context, two aspects seem to be relevant. Earlier the respective State High Courts exercised appellate jurisdiction over the SERCs. The new Act the transfers this appellate jurisdiction to a separate Appellate Tribunal. As one Tribunal will hear the appeals from the orders of different SERCs, certain uniformity will emerge over a period of time and this will help in settling the case law more easily. Transparency will be ensured since the proceedings of the Commission have to be conducted in a semi-judicial manner. As the operations of the SERCs will be entirely within the public domain (barring a few provisions covering commercial confidentiality), transparency is built into the system. The mechanism of Appellate Tribunal will help in reducing the possibilities of subjective judgements. The flip side of this framework may be that the regulators, beyond a point, cannot act as prime movers of the reform process.

As noted earlier, the new Act transfers substantial regulatory authority to the Commissions. The functions earlier exercised by the State governments, Central agencies, SEBs, Chief Electricity Inspectors and Commissioners of Electricity Duty are now vested in the Commission. The question naturally arises as to how they should go about their tasks. One has to remember in this context that the regulators have not only to operate within the framework laid down by the Act but also that they are still fledgling organizations. Regulators will not obviously get involved in day-to-day operational issues. It is envisaged that the goal setting for the operations and activities in the power system will be through regulations or codes. Technical entities are functioning in the sector since many decades. In such situations and matters a great deal of responsible technical self-governance is involved. Any interventions by the Commissions will have to be by way of exceptions. However, systems, procedures and monitoring mechanisms for periodically evaluating compliance with law, rules and regulations will have to be devised. The Commissions will have to develop Regulatory Information Systems and Response Mechanisms. Considering the nature of the tasks and the difficulty of assembling the needed experts in regulatory organizations, greater reliance will have to be placed on outsourcing such tasks to neutral, independent and professional agencies. It may not be possible to lay down a detailed uniform procedure covering all the Commissions.

## THE ACT IN THE CURRENT SCENE: MOMENTUM AND INERTIA

We have to constantly bear in mind that in regulatory situations we are not dealing with a *tabula rasa*. There are also existing agencies, personnel and methods which have been devised over a long period for regulating different segments of the power system. There are also codes, manuals, guidelines and standards of operating procedures. Before introducing changes, such existing systems and procedures have to be fully studied. Further the rationale of many currently operating systems may be implicit and not reduced to writing in any manual or code. Before formulating new regulations, these aspects need to be kept in view. One should aim at gradual change, though change also needs a critical initial momentum. The transitions have to be smooth so as to avoid needless commotion.

The Electricity Act 2003, being a forward-looking legislation, may have run quite ahead of the available infrastructure and system capabilities. Proposed innovations like open access and power trading will require many technical pre-conditions. There has to be enough capacity in transmission lines as well as inter-regional connectivity. Without these pre-requisites for moving energy from surplus to deficit areas, open access may not actually materialize. These considerations will also apply to open access in distribution systems. It has been observed that the binding constraints in power supply are at times in generation, at times in transmission, and at times in distribution. Similarly, difficulties will be faced by entities who want to move power into distribution areas from elsewhere. It may take a while to design and create technical preconditions on ground such as needed hardware, capacities, systems, codes, energy accounting methods and computerized monitoring devices for operating the provisions. Systems like electricity trading will need half-hourly or 15-minute real time data on movement of power through traders into distribution licensee areas and its assignment to various consumers. It is estimated that huge investments in transmission and distribution systems will have to be made (as the experience of UK shows) for effective introduction of open access system.

The new Act seeks to make the power sector more competitive. Here again the objective can be achieved only if the generating capacity exceeds the demand for power. This will in turn call for sizeable investments. Once adequate installed capacity is created, the distribution licensees will be able to source the cheapest power available during any significant period of the day.

Tariff determination and progressive reduction of cross-subsidization are important objectives in the new Act. As for tariff determination, it will have to be based on National Tariff Policy and this is yet to be finalized. One has to move away as far as possible from the cost-plus approach in tariff determination. In the medium term, however, in the absence of sufficient generating capacity, tariff determination will depend on regulations rather than on competition. Two considerations are important in this regard: delinking tariff through suitable mechanisms from rigid old Power Purchase Agreements (PPA) and prescribing a new set of financial and operational parameters for reducing power costs.[1] As is well known, interest rates have declined substantially and need to be appropriately factored into the financial norms. Augmenting technical and managerial efficiencies along with reduction of T&D losses can help lower costs.

An interesting issue arises about the delicensing of generation and removal of the earlier condition about techno-economic clearance of such projects by CEA. However, the Act lays down that the generation tariff will be determined by SERCs. Further, no distribution licensee can purchase power without clearance from the concerned SERCs. And without a PPA in place, it is virtually impossible for any power project developer to achieve financial closure. Such being the situation the delicensing of generation seems to be more *de jure* than *de facto*.

There are other issues which can arise in this context. In determining the generation tariff, SERCs will have to be guided by the National Tariff Policy and any operating norms that CERC may prescribe. But these may presuppose an approved capital cost since the financial norms are generally linked to a capital base. In the absence of an initial project cost approved by the Central or State government, or which is based on a techno-economic clearance of CEA, difficulties may arise. SERCs will have to judge the reasonableness of the initial capital expenditure with reference to acceptable benchmarks, norms and procedures. Since generation is delicensed, SERCs can take cognizance of the developers' process only at the stage when power is offered for sale to a distribution licensee. But by then the International Competitive Tender procedures would have been completed. Then the SERCs may have to undertake a separate retrospective due diligence procedure. One cannot offer any readymade solution at this stage. Rather our limited purpose is to flag the likely issues. The reduction of existing cross-subsidies for agricultural and domestic segments will take time. The level of such cross-subsidies is currently very high. If an attempt is made immediately to remove such cross-subsidies, the tariff level for agricultural and domestic sectors will shoot up. Accordingly, such revision will have to be phased over a period of time taking into account the overall situation prevailing in any state. Introduction of complete metering (especially covering agricultural sector) and reducing system losses can help in reducing the level of cross-subsidies. Systems like Availability Based Tariff (ABT) and time of use charges can have similar effects.

The new Act may have an inherent tendency to promote migration of such consumers of SEBs who bear the load of cross-subsidies. Provisions for open access for transmission and for movement of captive power can promote such migration by industrial consumers. This may in the short run add to the financial difficulties of SEBs. In view of the express provisions in the new Act, such resort to open access or for moving of captive power cannot be denied. However, as we noted earlier, the intervening transmission system should have enough redundancy for accommodating additional energy transfers. Although power may not be moved physically over long distances, constraints in the capacity of individual lines within a state may act as a deterrent. Thus, solutions will have to be found through new investments for augmenting system capacities. As captive units (especially small sets) may not be technically efficient (nor environmentally benign), SEBs will have to work towards better customer satisfaction to prevent migration of their high-paying flock.

The new Act envisages situations where captive power produced at one location will be transferred to factories at another location. However, the more common model is of captive power plants co-located within the battery limits along with production facilities. The definition of captive generation in the new Act is not free from ambiguity. On the one hand, the captive power has to be 'primarily' used by its generator. On the other hand it assumes that a significant portion of the captive power will be transmitted to diverse locations through diffusion of captive plant ownership. These issues will get resolved in due course based on actual practical experience.

## CONCLUSION

It is not possible to identify or anticipate all the issues which may emerge from a comprehensive legislation like the Electricity Act. There is no doubt that it seeks to address critical issues in the power sector. In particular, it makes a serious attempt to transfer critical decision-making functions to independent agencies. Thus, it separates policy making and regulations. Regulatory functions are separated from the day-to-day workings of the government. In this way, the new Act seeks to develop a structure and a procedure for creating an enabling environment for growth of the power sector. The new Act lays down the basis for rational decision-making with due regard to social objectives. It is necessary to recognize, however, that no legislation can by itself address all the issues in the power sector.

However, many of the solutions call for investments for augmenting systems capacity, technological upgradation, modern operating procedures, efficiency improvements, new managerial culture in SEBs or their successor entities, and responsible customer behaviour. Above all, the principal stakeholders will have to work together in a harmonious manner for creating an environment for overall improvement in the efficiency of the power system.

## NOTE

1. The N.K. Singh Committee has prescribed measures in this direction.

# 3

# *Markets through the Back Door*

SUDHA MAHALINGAM

## INTRODUCTION

The Electricity Act 2003 heralds the third phase in the process of power reforms underway in India since the early 1990s. The first phase was the IPP (Independent Power Producer) paradigm which ground to a halt within a few years of its launch owing to its many contradictions. The second phase was the half-hearted distribution reforms which failed to even take off the ground, except in two states, one a spectacular failure and the other still struggling to stay on course—with generous handouts from the government. The Electricity Act 2003 is an acknowledgement of the failure of the second reform phase—notably—privatization of distribution. With this law, the Indian government has now settled for selective privatization and competition, targeting the affluent consumer categories. The law is a confused and half-hearted attempt to rescue the Indian electric supply industry from the brink of bankruptcy. Its impact could well be the collapse of the already beleaguered SEBs with the poor and marginal consumers in both rural and urban areas bearing the brunt of the collapse.

## A DECADE OF TRIALS

Electricity reforms in India can be broadly categorized into three phases: the first IPP phase introduced in 1992 acknowledged power generation as a distinct commercial activity and invited private investors—both domestic and foreign—to invest in generation capacity, offering them extravagant incentives. Despite the initial euphoria over huge commercial opportunities for generators, the IPP model failed to take off owing to its many contradictions.[1]

Without formally abandoning the IPP model, yet another tentative reform phase was launched in 1996. This phase envisaged restructuring vertically-integrated utilities by unbundling them into their functional components, to be followed by privatization of distribution and generation. At the core of this model was the establishment of an independent regulatory commission to issue licenses and to fix bulk and retail supply tariffs. The regulator was envisaged as a mechanism to distance the government from electricity consumers and to act as a surrogate for electricity markets. It was anticipated that markets would emerge first in wholesale and eventually in retail electric supply.

Orissa in eastern India was the first state to be restructured thus, on a model devised, part-funded and overseen by the World Bank. The Orissa experiment was to be a precursor of similar restructuring in other Indian states. Despite the fact that the four distribution companies carved out of the Orissa State Electricity Board were successfully privatized, and all the reform milestones duly reached, it soon became apparent that the reform design was flawed. The experiment did not lead to anticipated outcomes. A variant of the Orissa model was then adopted to restructure the utility in the national capital Delhi and this model is being sustained with liberal subsidy support from the State government. Meanwhile, the unedifying outcome of the Orissa reform model and the requirement of continued state subsidy to sustain the Delhi model seem to have put paid to any hopes of pushing through the second reform phase in other Indian states. That this phase coincided with the compulsions of coalition governments at the Centre did not help matters either. State governments' response to the Centre's reform initiatives have often ranged from lukewarm to even outright resistance.

## THE THIRD PHASE OF REFORMS: ELECTRICITY ACT 2003 AND QUALITY OF SERVICE

The Electricity Act 2003 constitutes the third reform phase, and can perhaps be viewed as an acknowledgement of the failure of the second reform phase. While many states unbundled their electricity boards, privatization of distribution continued to remain an elusive target. Neither did the states display any eagerness to push ahead with distribution privatization, nor were there any viable private sector bidders willing to take over the difficult task of distribution, especially because it was viewed as a commercially unviable activity.[2] Meanwhile, the Central government, convinced as it was that privatization and introduction of competition were the only panacea for the ills of the industry, was anxious to somehow jump start the reform process. It resorted to shortcuts to markets through the concept of 'open access', which forms the centrepiece of the new legislation.

The Electricity Bill was first introduced in the Lok Sabha in August 2001, after having gone through seven revisions in draft form. It was then referred to a Parliamentary Standing Committee on Energy for detailed scrutiny. The Committee submitted its report in December 2002. The National Democratic Alliance (NDA) government then incorporated most of the recommendations of the Committee and the Bill was passed by the Lok Sabha in April 2003 and by the Rajya Sabha in May 2003. Presidential assent was obtained on 26 May 2003 and the Act came into force with effect from 10 June 2003. Subsequently, an amendment to the Act in December 2003 introduced certain changes. The new law consolidates the laws relating to generation, transmission, distribution, trading, regulation and use of electricity, and repeals the earlier laws in this regard.

At the core of the new law is the provision for 'non-discriminatory open access' to transmission and distribution networks—be they Central or State transmission utilities or licensees engaged in transmission and distribution, provided in Sections 38 to 42. By availing such open access, an Independent Power Producer (IPP), a Captive Power Producer (CPP) or even a distribution licensee of another area may elect to supply power directly to any consumer, bypassing the SEB or the regional distribution licensee as the case may be. Mandating open access would force the incumbent utility/licensee to turn its networks into something akin to public streets, to be used by anyone who wishes to transmit electricity from one point to another, albeit on payment of certain charges. In the event, it would be tantamount to introducing limited markets in electric supply. Here the term 'limited' is used to mean limited to those who elect to use this open access. The initiative for availing 'open access' could come from suppliers as well as consumers.

However, it might be reasonable to assume that only bulk consumers would be serviced by the suppliers since it would not be worth their (the suppliers') while to service a large number of small consumers. In industry parlance, this is referred to as 'cherry picking'. In fact, the proviso to Section 42 of the Act clearly limits open access to consumers requiring a minimum of 1 MW of supply. The competition, therefore, would be limited to bulk consumers requiring power to be supplied through high-tension and extra high-tension lines. It is also this category which pays some of the highest tariffs anywhere in the world, because high-tension consumers cross-subsidize other consumers. Switching suppliers might hold out prospects of cheaper power for this category of consumers. On the other hand, at least for now, low-tension consumers will have no incentive to switch suppliers even if they can, because, their tariffs are already low and are being cross-subsidized. In the event, the cumbersome task of servicing a large number of low-tension consumers—considered the unremunerative category—will remain with the incumbent licensee.

Why would IPPs rush to supply power directly to consumers? After all, IPPs have the comfort of long-term—often 30 years—Power Purchase Agreements (PPA) with the incumbent utility. In fact, these long-term contracts have even scuttled the prospects of a wholesale power pool by offering secure off-take. Even when they figure at the bottom of the merit-order dispatch system mandated by regulators owing to the high cost of their supply, IPPs' fixed costs are anyway payable by the utility that contracted to purchase their output. Thus, both the risk and reward elements that could have induced them to sell the bulk of their power directly to consumers by availing open access are indeed missing. However, in a situation where a State utility that has signed the PPA is unable to avail its full off-take—which is indeed the case with most SEBs—the IPP may find it worthwhile to sell the difference between the contracted and actual off-take directly to consumers who demand it, availing the open access now mandated by law. IPPs can also sell their surplus capacity over and above the specified off-take load factor directly to consumers. Similarly, CPPs would also be able to supply their surplus power to consumers in any part of the country, using the grid network for a fee. Even assuming a conservative estimate of about 20,000 MW of captive capacity in the country, it is likely that significant quantities of CPP surplus energy will enter the market for direct sale to bulk consumers. Enabling open access itself is an ideal incentive for setting up more captive units.

From the consumers' viewpoint, the incentive to switch from

existing supplier would mainly be the price and the quality of service. While new IPPs may be in a position to supply better quality power, since it will have to be wheeled through the existing transmission and distribution network, the quality of power that ultimately reaches the consumer is unlikely to be different from that supplied by the utility itself. In fact, once power is pumped into the grid by the IPP, it mingles with the rest of the electrons transmitted through the lines and becomes indistinguishable from the latter. Thus, the 'quality of service' argument is easily dispatched.

## THE THIRD PHASE OF REFORMS: ELECTRICITY ACT 2003 AND TARIFFS

Now we come to the price argument. The main argument in favour of open access has been the steep cross-subsidies borne by industrial consumers. As mentioned earlier, Indian industry pays very high tariffs and as such, there is a case for rationalization of tariffs to some extent. The preamble to Electricity Act 2003 clearly mentions 'rationalization of tariffs' as one of the objects of the law. Section 61(g) in Part VII also lays down the legislative intent in this regard. The section mandates that the regulator follow certain guidelines in fixing the tariffs. It reads: 'that the tariff progressively reflects the cost of supply of electricity, and also reduces and eliminates cross-subsidies within the period to be specified by the Appropriate Commission'.

The period was to be specified by the commissions within one year of the coming into force of the Electricity Act 2003, viz., June 2004. The amendment to the Act further provides 'open access shall be introduced within 5 years of the appointed date' namely, 9 June 2008. Thus, a deadline has been set to provide open access. In setting a deadline for introducing open access, the Centre has encroached upon the legislative sphere of the states.[3] While Article 251 of the Indian Constitution upholds the supremacy of a Central law over State laws in the event of an inconsistency between the two, it is not often that the Centre invokes this provision.[4]

It is also a moot point whether cross-subsidies can be or even should be completely eliminated in a country where just about 50 per cent of the households have access to electricity. Charging each category of consumers their actual cost of supply would mean the steepest tariffs for rural low-tension consumers—with least capacity to pay, while high-tension and extra high-tension consumers would pay the lowest tariffs.

Until cross-subsidies are eliminated, the regulator can allow open access only under certain conditions. Section 42(2) states,

> The State Commission shall introduce open access in such phases and subject to such conditions (including the cross-subsidies and other operational constraints) as may be specified within one year of the appointed date by it and in specifying the extent of open access in successive phases and in determining the charges for wheeling, it shall have due regard to all relevant factors including such cross-subsidies and other operational constraints.
>
> Provided that such open access may be allowed before the cross-subsidies are eliminated on payment of a surcharge in addition to the charges for wheeling as may be determined by the state commission.
>
> Provided further that such surcharge shall be utilised to meet the requirements of current level of cross-subsidy within the area of supply of the distribution licensee.

Thus, the onus of fixing the right quantum of surcharge that incentivises utilization of open access without jeopardizing the cross-subsidy is left to the commission. Certainly, it would make no sense for the regulator to fix a surcharge that would take care of the entire cross-subsidy element in the tariff. To the extent the surcharge is less than the cross-subsidy component, the incumbent utility's revenues would be affected. If a large number of high-tension consumers desert the utility in favour of direct purchase from IPPs/CPPs, the incumbent utility's already straitened finances could be seriously jeopardized, and might even lead to the collapse of the utility. While the objective of 'open access' is perhaps to act as a shock therapy to jolt the incumbent utility out of its torpor and induce it to rapidly reduce its technical and commercial losses, it might just end up killing the utility.

The Act enjoins upon the consumer who avails open access to pay the surcharge as well as the transmission charges, but is silent on how the losses along the open access line—whether technical or commercial—would be accounted for.

Even as it mandates that the tariff should progressively reflect the cost of supply and that cross-subsidies should be eliminated, Section 65 of the new law provides: 'If the State government requires the grant of any subsidy to any consumer or class of consumers in the tariff determined by the State Commissions under Section 62, the State government shall . . . pay in advance . . . the amount to compensate the person affected by the grant of subsidy.' It also adds a proviso that failure to pay the amount will render the State government's directive in this regard inoperative. While that might seem a reasonable proposition, in reality state governments' financial positions may not allow them to fork out the subsidy. The example of Andhra Pradesh government is a case in point. During the 2000-4 period it had directed the state regulator to lower the tariffs of certain categories of consumers undertaking to pay out subsidies to make up the difference, but owing to its financial constraints, had to settle for issuing bonds to the distribution companies and making book adjustments in lieu of cash. This provision of the Act is perhaps the most potent weapon available to the regulator to align tariffs with cost of supply.

Certain provisions of the new law purport to address consumer interests. Section 42(3) makes an almost apologetic effort to prescribe universal service obligations. It makes it obligatory for the incumbent distribution licensee to supply power on demand to every consumer within one month of application. The penalty for failure to do so has been fixed at a steep Rs.1,000 per day. However, the penalty is applicable only in cases where electric lines and, sub-stations already exist, not in areas where there is no connectivity yet. However, in order to compensate the distribution licensee for fixed costs arising out of its obligation to supply, Section 42(4) provides for levy of additional surcharge to be determined by the appropriate commission.

Section 42(6) provides for an ombudsman to be designated or appointed by the State Commission to redress consumer grievances. An electricity consumer in this regard would be the same as defined in Section 2(1)(d) of Consumer Protection Act 1986. Section 61(d) requires the regulator, while fixing tariffs, to be guided by consumers' interests. Section 57 provides that the regulator may set standards of performance for the licensee to protect the interests of consumers.

Power trading has been recognized as an activity that requires licensing. Section 52 empowers the regulator to lay down capital adequacy for undertaking this activity. Incumbent distribution licensees can engage in power trading without a licence.

In order to ensure the independence of the regulatory commissions, the Act provides for the establishment of a separate fund. All grants and loans provided by the State government to the commission, all fees received by the commission and any other receipts would go to make up the fund which will pay the salaries and allowances of the commissioners and meet the expenses of the commission. This should, in fact, go a long way in freeing the commissions from financial dependence on state governments.

The Act also provides for the establishment of an Appellate Tribunal to hear appeals against the orders of regulatory commissions. Sections 110 to 125 deal with the constitution, scope, powers and functions of this authority which is bound to decide the cases brought to it within 180 days. Appeals against the orders of the Authority lie with the Supreme Court.

A signal contribution of the new law is the delicensing of generation—be it IPPs or CPPs. However, hydroelectric projects involving specified capital costs will have to be cleared by the Central Electricity Authority, under Section 7 of the Act. In all probability, mini-generation plants—perhaps diesel gensets—will proliferate in urban residential colonies, because Section 13 authorizes the SERCs to exempt users' associations, panchayat institutions, coöperative societies, non-governmental organizations and franchisees from procuring licence for power supply. Besides its impact on the viability of the SEBs, proliferation of captive generation could lead to unplanned increases in capacity, jeopardizing optimal fuel and technology choices. The Parliamentary Standing Committee that scrutinized the Bill had desired that delicensing should be accompanied by provisions to ensure that new plants were established in accordance with the National Plan, using appropriate technology and fuels, and without providing grounds for dumping outdated technology. But its recommendations have not been incorporated into the Act.

One important provision that did not figure in the original Bill but found its way into the Act is a proviso in Section 14, which stipulates that 'the Appropriate Commission may grant a licence to two or more persons for distribution of electricity through their own distribution system within the same area', if such licensee complies with the requirements of capital adequacy, creditworthiness and so on. It further states that no such applicant will be refused a licence on the grounds that a licensee already exists in the same area for the same purpose. This is tantamount to an allowance for duplication of networks, a waste of national resources that even the most prosperous economies refrain from advocating.

It is now an acknowledged fact that at the root of the ills of the power sector in the country is the massive commercial and technical losses. In a situation where less than half of the electricity that is supplied is paid for, tackling commercial and technical losses would raise the revenues of the utilities and further tariff increases may not even be required. However, there seems to be little thrust in that direction in the present law. Section 135 of the Act prescribes a penalty of three times the financial gains accruing from theft for the first conviction and only imprisonment for subsequent convictions, that too for theft of more than 10 KW hours. Provisions have been included to empower authorities to enter any premises and search and seize if there is reasonable cause to believe that electricity is being stolen or meters are being tampered with, but these measures are being decried by civil society as encroachment on civil rights. The stipulation on time-bound 'universal metering' contained in Section 55 is considered too ambitious and perhaps infeasible.

The emphasis on cost-reflective tariffs based on commercial principles enshrined in Section 61(b) and elimination of subsidies in the absence of stringent anti-theft laws is a sure recipe for pushing the marginal and poor consumer segments out of the grid, limiting electricity supply to those who can afford it. The signal achievement of the new law is that it has commodified electricity and entrusted its supply to the markets. In a situation of supply shortages, market manipulation by generators could seriously impede the functioning of even this limited market.

## CONCLUSION

The Electricity Act 2003 is an acknowledgement of the government's failure to privatize SEBs. With this law, plans for wholesale privatization of distribution seem to have been abandoned in favour of a short-sighted measure—of selective privatization and competition targeting the affluent consumer categories. The law has also failed to grasp the main weakness of India's electric supply industry, namely the large-scale commercial and technical losses. It is rather lenient on power thefts and displays confusion when it advocates commercial principles for tariff fixation without a corresponding emphasis on elimination of thefts. Such an approach, coupled with the stipulation to charge cost-reflective tariffs and eliminate cross-subsidies could well exclude the poor and marginal consumers in both rural and urban areas from access to electricity. In the event, the provisions relating to 'safeguarding consumer interests' will remain a mere platitude.

## NOTES

1. For a detailed critique of the IPP paradigm, see Sudha Mahalingam, 'Economic Reforms, Power Sector & Corruption', in *The Political Economy of Reforms in India,* Jos Mooij (ed.), Sage, forthcoming.
2. The Montek Singh Ahluwalia Committee estimated that only 46 per cent of the power supplied was paid for, the rest being accounted for by technical and commercial losses. See Government of India, Ministry of Power, 'Report of the Expert Group on Settlement of SEB Dues', May 2001.
3. Similarly, Section 82(1) makes it mandatory for the states to establish a regulatory commission within six months of the appointed date (6 June 2003).
4. At the time of writing, the United Progressive Alliance government that had assumed office at the Centre was reportedly considering amendments to the Electricity Act 2003. The pressures of coalition politics indicate that the deadlines set by the Act might come up for amendment.

# 4

# *Benchmarking of Electricity Distribution Companies in India*

PREM K. KALRA, V.P. SINGH AND
YOGESH K. BICHPURIYA

As the pace of change accelerates in the twenty-first century as a result of technological opportunities, liberalization of world markets, and demands for innovation, quality and speed, organizations have to readjust and realign their operations to counter these challenges. The pace of change has increasingly forced organizations to be more outward looking, market-oriented and knowledge driven. A useful tool that can help businesses build strong capabilities, ensure inward flow of ideas and establish true competitive gaps is benchmarking.

Deregulation and the privatization of electric companies are taking place at a rapid pace in India. With the emergence of a large number of distribution companies, there is a need for assessing their comparative efficiency to promote competition. This paper proposes a methodology for the efficiency benchmarking of distribution companies in a typical Indian system.

## INTRODUCTION

The poor financial and technical condition of State Electricity Boards (SEB) in India is well known. Due to the cost inefficiency of the existing power plants, the SEBs are not interested in increasing capacity to meet increasing demand. The losses are very high on the distribution side. These losses include technical as well as non-technical. Non-technical or commercial losses may involve theft of power, unmetered use of power and poor billing. All these factors contribute to poor collection efficiency of SEBs. Therefore, SEBs are avoiding to invest more as they are not getting adequate returns. This results in frequent power cuts and poor quality of supply. Besides, there is extreme variation in the tariff structure for various types of consumers. While agricultural and domestic consumers get the benefit of subsidies, the industrial consumers are charged a high tariff.

The power sector in India faced major problems like: (*a*) breaks in the payment chain, leading to multiple inefficiencies in the supply chain; (*b*) tariffs not coordinated according to different consumer categories; (*c*) blackouts and (*d*) poor power quality. It was recognized that meaningful reforms would not take place unless more distribution areas were privatized. This has been encouraged by the Electricity Act 2003, passed by the Indian government. The Act makes it mandatory for states to set up electricity regulatory commissions.[1] Apart from making power trading a distinct activity, the Act also allows open access to transmission and distribution systems and rationalization of tariffs. This restructuring typically involves moving from an integrated monopoly-utility structure to competitive industry structure and a shift from government/public owned utilities to private utilities.

The purpose of the Electricity Act 2003 is to create a competitive market for electricity distribution. Distribution is now seen as a three-element business:

(*i*) Network business—ownership and maintenance of the network.
(*ii*) Power market business—supplying power to consumers, either from one's own source or from a third party.
(*iii*) Services business—monitoring, billing and collections.

The privatization of the distribution sector is an important need of the power sector reforms. The previous reforms also focused mainly on the distribution side of the power sector. For a given set of input resources the efficiency of any system can be maximized by maximizing the output. For a developing country like India where the number of resources for electricity generation is limited, the reforms in the distribution side have more importance.

As the process of deregulation is looked at as a reform in the power sector in India, the increasing number of private distribution companies should provide better and reliable supply to consumers.

To ensure quality of service and to maximize the use of given resources, all the licensee electricity utilities are required to prepare and publish their annual financial data as per the provisions of Schedule VI of the Electricity Supply Act 1948, and Indian Electricity Rules. These reports should have all the information about the performance of the utilities.

Measuring the performance and efficiency of various distribution companies is not a simple task. The performance of distribution utilities depends on various factors such as quality of supply, tariff structure, capital cost, operation and maintenance cost, and losses. There is a need to understand the Indian electricity network and then to develop a methodology which can judge the operational efficiencies of the distribution companies. Thus, there is a need for *benchmarking* the distribution companies.[2]

Benchmarking is a collaborative learning process among a group of companies to focus on specific operating practices, compare measures and results and identify improved processes within the organizations. In the power sector, benchmarking is done amongst the distribution companies who do similar things, but have developed their approach in different ways. The determination of benchmarks allows one to make a direct comparison. Any identified gaps are improvement areas.

In other words, benchmarking is a method to measure the performance and the potential efficiency of organizations. By comparing the operating efficiency of different distribution companies within a country, further improvements in the model can be made.

As the number of distribution companies is increasing to meet the increasing demand of customers, quality of service has received much attention. The benchmark for quality of service enables distribution companies to get information about the quality of service provided by them and to know what the customers' quality needs are.

## PRODUCTION PROCESS

Electricity distributors use their network and transformer capacity, together with labour inputs, to deliver a certain number of electrical units to a specified set of consumers and to meet a level of peak demand. At present, practical implementation of efficiency benchmarking of the distribution companies is carried out in countries such as USA and the Netherlands.[3] The variables chosen as input and output for evaluating the efficiency of electricity distribution companies in England and Wales, Sweden, the US and New Zealand are shown in Fig. 4.1.

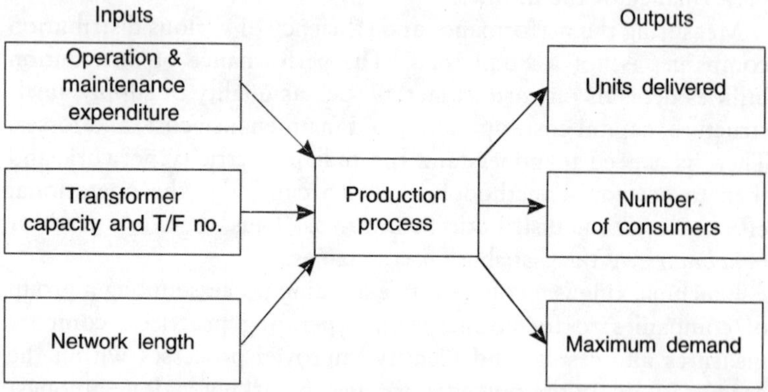


FIGURE 4.1: ELECTRICITY DISTRIBUTION PRODUCTION PROCESS

The experiences gained by these countries cannot be exploited in India as such because the parameters on which the efficiency calculations are based are different in each country.[4] One typical feature of the Indian electricity market is that the distribution companies vary greatly from each other in terms of size and operational environment, for example, the length of the distribution networks as shown in Fig. 4.2.

Therefore, an independent strategy for efficiency benchmarking of the Indian distribution companies is essential.

To model the relative efficiency of distributors, no matter what

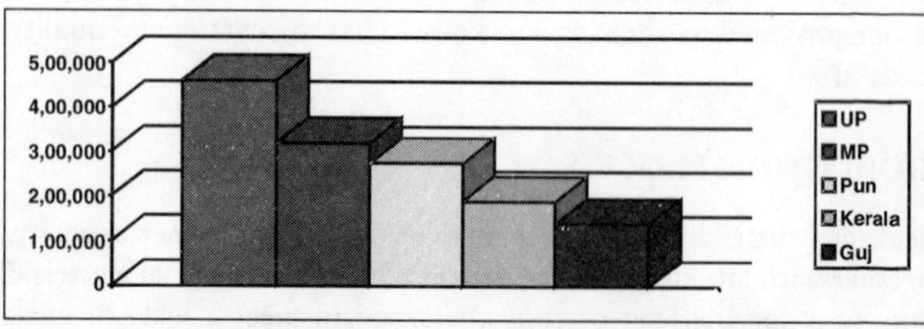


FIGURE 4.2: NETWORK LENGTH OF DIFFERENT STATES

type of services they may produce, it is necessary to first define a production relationship or *production process* that is representative of the industry being examined. Distributors consider a wide range of inputs and outputs to provide services to consumers. The choice of inputs and outputs used for benchmarking the distribution companies varies from place to place and largely depends upon management practices and operating environment. Similarly, the nature of services provided by distributors varies according to the nature of consumer demands. Thus, the proposed benchmarking methodology is capable of modelling the interaction between multiple inputs and outputs for efficiency evaluation which is well suited for the Indian system. Thus, the variables chosen as inputs and outputs for the proposed model as shown in Table 4.1 varies that from the other tested model of different countries as shown in Table 4.2.

TABLE 4.1: PROPOSED PRODUCTION PROCESS MODEL

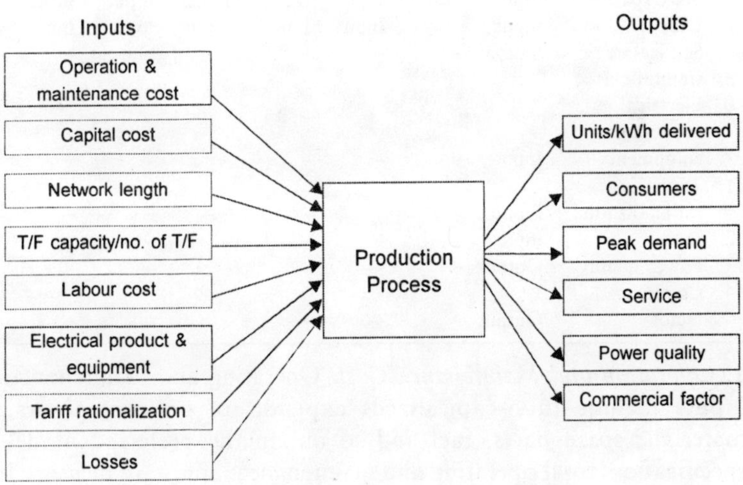


## INPUT VARIABLES

The inputs selected in the proposed production process model consist of those parameters which directly affect the infrastructure and working culture of the company. They can be classified into the following categories:

TABLE 4.2: COMPARISON OF THE PROPOSED MODEL WITH DIFFERENT EXISTING MODELS

| S. No. | Variable | Proposed Model for India | DTe (2000) | Hijalmarsson & Vaiderpass 1992 Sweden | Weyman Jones 1992 England & Wales | Weyman Jones and Burns 1994 England & Wales | Pollitt 1995 England & Wales, USA | London economics 1999 England & Wales, USA, Australia New Zealand |
|---|---|---|---|---|---|---|---|---|
| 1. | Operating expenditure | Input | Input | – | – | – | – | – |
| 2. | Units consumed | Output | Output | Output | Output | Output | Output | Output |
| 3. | Consumers | Output | Output | Output | Output | Output | Output | Output |
| 4. | Peak demand | Output | Output | – | Output | Output | Output | Output |
| 5. | Network length | Input | Output | Input | Input | Input | Input | Input |
| 6. | Transformer no./Capacity | Input | Output | Input | Input | Input | Input | Input |
| 7. | Labour | Input | – | Input | Input | Input | Input | Input |
| 8. | Service area | Output | – | – | – | – | Output | – |
| 9. | Capital cost | | | | | | | |
| 10. | Electrical product and equipment | Input | – | – | – | – | – | – |
| 11. | Tariff rationalization | Input | – | – | – | – | – | – |
| 12. | Losses | Input | – | – | – | – | – | – |
| 13. | Power quality | Output | – | – | – | – | – | – |
| 14. | Commercial factor | Output | – | – | – | – | – | – |

1. *Operation and Maintenance Cost:* Operating and maintenance inputs include (non-capitalized) expenditure on contractors, materials, spare parts, fuel and so on. In the preferred model specification, total operating and maintenance inputs are measured by annual expenditure rather than as a physical measure, given the very diverse nature of these inputs.

2. *Capital Cost:* Capital inputs represent a significant proportion of inputs utilized to provide energy distribution services (on a cost basis). The stock of capital can be measured in either physical or monetary terms. The stock of capital inputs that is used by distributors to provide energy delivery typically includes:

(*a*) Poles and mains;
(*b*) Transformers;
(*c*) Buildings; and
(*d*) Vehicles and other equipment.

3. *Network Length:* There are a few countries like the Netherlands which considers network length as output. However, considering so will undermine the assessment of the ability of the distributors to improve efficiency. The reasons for this are outlined below:

(*a*) A network (poles, wires and transformers) is required for a distributor to deliver the fundamental output energy when the consumer requires it.
(*b*) A network results in costs. Inputs are normally specified in terms of costs whereby outputs are normally related to the goods and services for which consumers are prepared to pay an amount to receive.
(*c*) Specifying the network as an output assumes that managers have no control over whether it is provided more or less efficiently.

In the proposed model, network length is taken as input; further, network length measurement is done on the circuit kilometres, (length between poles by number of circuits) which is in contrast with the other country models such as USA and New South Wales where distributors consider route kilometres (the linear distance between poles regardless of how many circuits are supported).

The network length can be sub-divided into two parts:

(*i*) Overhead line
(*ii*) Underground line

4. *Transformer Capacity/Number of Transformers:* Different distribution companies can be compared by the transformer capacity/ number of transformers used by them to meet the required supply demand while maintaining the quality.

5. *Labour Cost:* In the proposed model labour is considered as a separate input quantity. This is done in order to provide a more detailed information about the potential sources of inefficiency.

6. *Electrical Product and Equipment:* The electrical product and equipment used by the company and consumers also need to be

benchmarked, which is not considered in any of the existing models of benchmarking the distribution companies. The products used by the company and the consumer, such as transformers, switching devices and home appliances are sources of magnetic field which may exceed permissible limit and hence affect health in the long term. Also, the harmonics and high current due to low power factor devices would increase distribution losses drastically. Further it would also contribute to reducing the life of equipment and hence value of money would be possible for supplier. It is proposed that benchmarking design specification and performance levels are must for improving the long-term health of both utility and consumer.

7. *Tariff Rationalization:* Rationalization of tariff structure will lead to better supply side management along with demand side management. This would ensure that available quality power reaches maximum consumers. Further incentivizing collection through profit centres would lead to better collection efficiency and good returns to utility. In India, for example, electricity supply to the agricultural sector is highly subsidized. Thus, according to the Indian system each class of consumers like industry, domestic and agriculture must have different benchmarks.

8. *Losses:* Losses have been included as an input because in India there is a large amount of distribution losses compared to countries such as the USA. Also, there is a huge difference amongst the company distribution losses. There are two types of losses which are considered in this model:

(*i*) Technical losses
(*ii*) Commercial losses

Electricity losses should have been included as an input; if no data is available at the distribution level then the approximation according to the company tariff order can be taken.

### Output Variables

The outputs of the proposed production process model are those which greatly affect or are affected by the consumer. They are classified as follows:

1. *Units/kWh Delivered:* Electricity distribution is about delivering energy from the high voltage network to the low voltage consumer's

site. Two aspects of energy in the distribution system should be considered:

(a) The absolute level of energy that passes through the system in a period of time; and
(b) The duration of peak load.

The amount of energy delivered is important to the specification of the production process for electricity distribution as:

(i) It reflects the degree to which the distributor is meeting his objective of facilitating the flow of energy to consumers;
(ii) Combined with the number of consumers, it indicates relative energy density; and
(iii) Along with peak demand, it accounts for relative capacity utilization and consumer mix.

2. *Consumers:* India is the second largest country in the world in terms of population (over 1 billion). Thus, while benchmarking consumers as an output variable in the Indian context, it has to be subdivided into two categories:

(*i*) *Type of Consumers:* In this section we classify the consumers according to the amount of their electricity consumption and their operating voltage level.
   (a) Small
   (b) Large
   (c) Domestic
   (d) Industrial
   (e) Agriculture
(*ii*) *Number of the Consumers:* Consumer number reflects the quantity of low voltage lines that a distributor requires to connect his consumers, that is, the lines it must construct and service. Consumer numbers act as a proxy for the number of connection points for each distributor. It is therefore important that these are included as an output for the distributors. The ratio of consumers to kWh accounts for energy density.

3. *Peak Demand:* Incidental peak demand provides a proxy for the network and transformer capacity required by the distributor to allow the delivery of energy to consumers at peak times. This reflects

the fact that the distribution system must be constructed to meet peak demand at its incidence, even if this is well above average demand. The inclusion of incidental peak demand ensures that a distributor who requires more inputs to meet relatively high demand is not penalized in the efficiency assessment.

4. *Service:* Efficiency of the distribution company is judged by the service it provides to its consumer and hence the revenue collected by it. Thus it consists of two parts:

(*i*) *Revenue Collection:* Any distribution company in order to have better infrastructure must have regular returns and maximum profit by implementing a better revenue collection policy. Thus it is a managerial process and varies from company to company.

(*ii*) *Area Serviced:* Companies must also be compared by the amount of area that they are serving because it may be so that they are covering a very large area in which the population density is very less; for example in the hills hence the consumer serviced to the area covered will be low. On the other hand, a company doing business in a metropolitan city may be covering a small area but he is covering a very high population density. Hence, their consumer serviced to the area covered will be high and hence should not be compared with the company doing business in a low population density area.

5. *Power Quality:* Power quality of the energy supplied to the consumers is an important variable which distinguishes distribution companies according to the following parameters:

(*i*) *Power Factor:* Low power factor of distribution products leads to more loss of power. Hence the products used by the consumer should be benchmarked by the company to give them an incentive for using high power factor products.

(*ii*) *Availability and Reliability:* Availability is the time in hours the transmission system is capable to transmit electricity at its rated voltage from the supply point to the delivery point and it will be expressed in annual term. Reliability benchmarking should attempt to quantify the effect of factors such as lightning strokes which affect normal operating conditions in the company service territory. Thus it helps in assessing the real performance of the company in providing reliable service.

(*iii*) *Interruption Time:* In India due to shortage of generated power there are frequent power cuts which at times even exceeds to 24 hours or more. Thus, interruption time is an important parameter to be taken into consideration.

6. *Commercial Factor:* A company is judged by the value of its share in the market. This largely depends upon the number of consumers that are the part of that company and feel secured in investing their money in that company.

## Efficiency Evaluation

Efficiency of the model can be assessed either on an input minimizing or output maximizing basis. An input minimizing orientation assumes that the output level is fixed and measures efficiency on the basis of minimizing the amount of inputs used to provide that output. An output maximizing orientation treats the inputs as fixed and measures efficiency on the basis of the distributor's ability to provide as much output as possible with a given level of inputs.

In India there is a huge gap between supply and demand. India has a fixed amount of resources such as generating capacity, lack of funds and so on. Keeping this constraint in view we try to maximize our output so as to provide better quality service to as many consumers as possible, rather than keeping the demand fixed and minimizing the amount of resources. Thus, in the proposed model, output maximizing orientation should be used.

There are a few unavoidable events such as earthquake and storm which lead to a long interruption in power supply to the consumers. Thus, the quantity and quality of the power supplied to the consumers during that period should not be considered while calculating the output efficiency of the companies.

The deregulation and privatization of the power sector are underway in India. With the introduction of the Electricity Act 2003 there is more freedom of open access to generation, transmission and distribution. With more and more active participation of private companies in the field of electricity distribution there is a need to assess their efficiency. This is important so to promote healthy competition amongst the companies with an intention of providing maximum benefit to the consumers.

Benchmarking is widely regarded as a 'corporate miracle' and has become popular among practitioners of various disciplines.

Benchmarking of distribution companies has been done in few countries like the USA and England and Wales. The experience gained from them promoted to use benchmarking for the distribution companies in India. But the production process model used by them cannot be used as such for doing qualitative and quantitative study and comparison of distribution companies in India. For example, the model implemented in USA does not consider interruption time, losses and availability, whereas these factors have a strong impact on the socio-economic development and hence are considered in the proposed model for the Indian system.

## CONCLUSION

Benchmarking of the electricity distribution companies in India has much importance. The planning of capital investment, load growth, better quality of supply should be covered in an in-depth study of performance of these distribution companies. Benchmarking for losses, capital cost and quality of service should be established.

At present only the proposed production process model well suited for the Indian system is introduced. The ongoing and future work aims at doing a detailed case study, whose success will lead to its practical implementation. With the successful practical implementation of this, distribution companies operating in India will not only be able to increase their efficiency but will also link themselves to world-class performers in the power sector.

## NOTES

1. The Electricity Act 2003, Ministry of Law and Justice, Government of India.
2. 'What is benchmarking', American Productivity and Quality Centre Report, USA, 1997.
3. Nillesen P. and J. Telling, 'Benchmarking Distribution Companies', *EPRM Electricity*, March 2001; J. Lassila, S. Viljainen and J. Partanen, 'Analysis of the Benchmarking Results of the Electricity Distribution Companies in Finland', IEEE Postgraduate Conference in Budapest, Hungary, 11-14 August 2002.
4. 'Efficiency and Benchmarking Study of the NSW Distribution Businesses', *London Economics*, Report Commissioned by the Independent Pricing and Regulatory Tribunal, New South Wales, Australia, February 1999, www.ipart.nsw.gov.au. ; Lawrence Kaufmann, Mark Newtor Lowry and David Hovde, 'CitiPower Performance: Results from International Benchmarking', Pacific Economic Group, Madison, Wisconsin, USA. http://www.esc.vic.gov.au/docs/electric/pegciti.pdf.

# PART II

# TECHNICAL CHOICES AND MANAGEMENT

# 5

# *Need for Distribution System Reforms and Customer Training*

A.K. SAXENA AND PREM K. KALRA

Restructuring and deregulation has led to greater competition in the power sector. The distribution system has also been opened for private participation and open access. These changes will bring an improvement in the existing state of the power distribution system. The responsibility of the distribution companies is likely to be increased. The customers too have to be careful and have to strictly monitor their energy consumption pattern. Once the reforms in the distribution sector come into effect, the need for proper training of customers will become essential. This paper presents brief aspects of the need for distribution system reforms and customer training.

## INTRODUCTION

The electrical power distribution system is geographically very widely distributed. Increase in the size and complexity of this system have gradually changed the requirements for control and monitoring of system parameters. The number of electricity consumers is growing. However, the utilities are paying more attention towards reforms in the area of power generation and transmission. The customers connected to the distribution network need continuity in power supply and good quality of power. More sophisticated electrical/electronic gadgets are flooding the market, and these require good power quality. Good power quality essentially means proper voltage level.

Rapid growth in the demand for electricity is expected throughout the world. Many nations including India are restructuring their power industries to increase efficiency and cut down costs. A competitive environment for the electricity market has been created and this will affect the structure and ownership of the supply industry. The distribution sector is now opened for private participation and this will encourage competition. Each party will try to attract customers and will have to maintain the quality of supply. We need efficient systems to ensure that the voltage profile and continuity of power are maintained at acceptable values. The security and economy of operations are the other aspects.

If the distribution sector reform ensures good quality of electrical power, then the customers will be encouraged to use maximum electric power. An unwise decision by the customer to use maximum power without optimizing the use of electricity will cause a burden on him. Customers are not expected to shut off their load to save money, but we can teach them the means to save power by selecting a suitable quality of electrical gadgets. Most Indian industries, in order to reduce cost, produce inferior quality products. These products are not energy efficient and consume more power without giving the desired output. So customers must be trained to choose and use suitable gadgets for a particular operation.

To ensure reliable and continuous supply from the distribution sub-station, and, efficient utilization of the power at the customer end, we must emphasize and plan suitable measures for maintaining the distribution system and providing training to the customers. Several authors have presented ideas on load management and operator training, but none have suggested measures for end-user training. Before designing a suitable training programme, a survey of customers at different locations needs to be done to decide about the curriculum and action plan of training.

This paper highlights the various aspects of customer training. The arguments are supplemented by various test results.

## DISTRIBUTION SYSTEM REFORMS AND CUSTOMER TRAINING

The responsibility for continuous supply of power to various customers at different locations rests on the persons operating and maintaining the system. The load on the system is increasing day by day but the system is not upgraded at the same pace. This causes frequent tripping. The distribution lines become overloaded and supply power beyond their safety limit. This causes increased distribution losses. Even if we neglect the losses due to theft, the other technical loss due to improper conductor size and unbalanced load is a very huge amount. The losses due to harmonics introduced into the system by modern electrical gadgets are not accounted for so far. The distribution system load is also not balanced. Customers switch connections according their own will. If one phase goes off he switches to another phase. The distribution system at the consumers' premises is mostly maintained by private linemen as there is an acute shortage of operating personnel with the concerned department.

The distribution system reforms include the following aspects of system operation: sub-station metering, load balancing, line maintenance, line patrolling, feeder rescheduling/sectionalizing (in case of faults) and planning, capacitance switching for voltage control, sub-station maintenance, energy auditing, disturbance analysis, voltage level control by OLTC, maintenance of proper staff for repair. All these are necessary to ensure quality power.

The electricity consumption habits of Indians are seriously deficient. They are normally not aware of the qualities of the products available in the market. They do not know what will be the optimum requirement for their need and how and where to install a particular gadget. A person spends more than is necessary for shelter, clothing and food. Futhermore they are susceptible to the enticement of false advertisements. It is the business of education to correct and train them.

For customer training we must first classify the customers into various user categories. This classification can be either on the basis of income group or on the basis of educational qualifications. A further sub-classification can be done on the basis of the type of the electrical gadgets they use in their day-to-day life. Before designing a training curriculum we must first pay attention to the following questions:

Why give the training? What should be taught (activities, attitude)? Where should the training be conducted (at consumers' premises/training centres)? With what equipment and facilities (how many units, special training manuals)? When shall the training be given and for how long (one week/month)? In what sequence will the training be given? How are activities performed? How will the training material be used? Who shall teach during the training? To whom shall the training be given? What should be the method of evaluation? How to monitor the difficulties/errors/methods of performance? What was the attitude of the trainee?

Customer education must deal with the elimination of waste in energy consumption, improvement of standards and minimizing financial problems. In this paper we have tested various electrical bulbs, compact fluorescent lamps (CFL), fans, stabilizers and water pumps for their performance at different voltage levels and the comparison is presented. By these tests we can convince the customer about the selection of a particular item not only on the basis of cost but also quality and performance. The net saving in the normal life span of the equipment can also be predicted on the basis of these tests.

## THE CORE OF TRAINING: MARGINS FOR EFFICIENCY

### TUBE LIGHTS AND CFL

We are using 17-20 per cent of power generated for lighting compared to the developed countries' average of 8-9 per cent. Average light levels in major installations even at this high usage is 40 to 50 per cent lower than those in developed countries which are reaping the benefits of advanced technologies and best maintenance practices.

Thus, there is scope for energy conservation using efficient lighting technologies. At the present level of aggregated losses, saving one unit of electricity at point of use saves two units at point of generation.

Energy-efficient slim tube lights not only use minimum of natural resources but coupled with high-efficiency electronics ballasts offer the best choice for energy-efficient indoor lighting. While deciding about the selection of lighting equipment for energy conservation always consider the total system of power and calculate running as well as maintenance costs, taking into account savings on account of reduced air-conditioning load while recommending heat saving luminaries using modern electronics ballast and energy-efficient light sources. Once installed it is equally important to maintain any luminaire as deposition of dust on lamp and parts of luminaire does reduce downward light output ratio. Also, instead of waiting for lamps to fail, lamps should be periodically replaced according to manufacturer's data as lumen descent after claimed lamp life will generally be higher. Many of the old technology products, even if you put zero value on their capital costs, would be bleeding you steadily by higher running cost.

Now electronic chokes are available which are cheaper and lighter in weight as compared to copper choke (ballast). To produce the same type of light with electronic ballast as compared to magnetic ballast, we require 10 per cent lower power to the lamp. This coupled with the fact that the ballast Watt loss for electronics ballasts is only of the order of 1.5-3.0 W, results in overall system efficiency improved by 30 per cent. Table 5.1 shows the characteristics of different CFLs.

TABLE 5.1: CHARACTERISTICS OF VARIOUS CFLS

| Voltage (Volts) | A | | | B | | | C | | |
|---|---|---|---|---|---|---|---|---|---|
| | Current (Amps) | Power (Watts) | LUX | Current (Amps) | Power (Watts) | LUX | Current (Amps) | Power (Watts) | LUX |
| 150 | 0.01 | 8 | 86 | 0.1 | 8 | 168 | 0.1 | 8 | 153 |
| 160 | 0.01 | 8 | 94 | 0.1 | 8.25 | 193 | 0.1 | 8 | 173 |
| 170 | 0.01 | 8 | 102 | 0.11 | 9 | 211 | 0.1 | 9 | 188 |
| 180 | 0.015 | 8 | 108 | 0.11 | 9 | 229 | 0.1 | 9 | 204 |
| 190 | 0.015 | 8 | 116 | 0.11 | 11 | 244 | 0.1 | 9 | 220 |
| 200 | 0.018 | 8.5 | 123 | 0.11 | 12 | 259 | 0.1 | 9 | 233 |
| 210 | 0.02 | 8.5 | 129 | 0.11 | 12 | 274 | 0.1 | 11 | 249 |
| 220 | 0.04 | 9 | 135 | 0.11 | 13 | 288 | 0.1 | 12 | 262 |
| 230 | 0.05 | 10 | 140 | 0.11 | 14 | 305 | 0.1 | 12 | 275 |
| 240 | 0.05 | 12 | 147 | 0.12 | 15 | 317 | 0.1 | 13 | 287 |
| 250 | 0.05 | 12 | 152 | 0.12 | 16 | 330 | 0.1 | 14 | 300 |

## Incandescent Lamps

Normally, 60 W and 100 W incandescent lamps are used for domestic lighting. They are available under different brand names and at differing prices. Table 5.2 shows the performance of different bulbs. The customer must be trained to calculate the net saving in energy cost while ensuring the desired level of lighting and select suitable quality of bulbs. Table 5.2 indicates the effectiveness of the individual lamps.

TABLE 5.2: EFFECTIVENESS OF LAMPS

| Voltage (Volts) | A @ Rs. 9 | | | B @ Rs. 9 | | | C @ Rs. 5 | | |
|---|---|---|---|---|---|---|---|---|---|
| | Current (Amps) | LUX | Power (Watts) | Current (Amps) | LUX | Power (Watts) | Current (Amps) | LUX | Power (Watts) |
| 150 | 0.35 | 47 | 299 | 0.32 | 50 | 209 | 0.30 | 43 | 178 |
| 160 | 0.36 | 60 | 362 | 0.33 | 56 | 265 | 0.32 | 52 | 215 |
| 170 | 0.37 | 64 | 433 | 0.35 | 60 | 330 | 0.325 | 56 | 268 |
| 180 | 0.39 | 72 | 542 | 0.36 | 67 | 402 | 0.33 | 61 | 327 |
| 190 | 0.40 | 80 | 650 | 0.37 | 72 | 480 | 0.34 | 68 | 404 |
| 200 | 0.41 | 85 | 762 | 0.38 | 80 | 580 | 0.35 | 73 | 475 |
| 210 | 0.42 | 92 | 890 | 0.39 | 84 | 690 | 0.36 | 80 | 566 |
| 220 | 0.43 | 97 | 1044 | 0.40 | 92 | 789 | 0.37 | 85 | 668 |
| 230 | 0.44 | 104 | 1195 | 0.41 | 96 | 890 | 0.38 | 89 | 750 |
| 240 | 0.46 | 109 | 1358 | 0.42 | 102 | 1050 | 0.39 | 96 | 862 |
| 250 | 0.47 | 117 | 1537 | 0.43 | 108 | 1202 | 0.40 | 102 | 977 |

## CEILING FANS

Ceiling fans are available in different sizes and of different power ratings. Normally, 1,200 mm and 1,400 mm size of fans are available. The customers are not able to decide that for their room size which fan will provide better air. These fans consume different power depending on their size. The air delivery depends on the angle of blades; angle of blades affects power consumption also. So, customers need training to select fans of proper size and proper quality. Table 5.3 shows the average power consumption (Watt) of some of the samples.

TABLE 5.3: EFFECTIVENESS OF CEILING FANS

| Voltage (Volts) | Power (Watts) | | |
|---|---|---|---|
| | A | B | C |
| 200 | 96 | 60 | 80 |
| 210 | 104 | 62 | 84 |
| 220 | 112 | 64 | 86 |
| 230 | 120 | 66 | 88 |
| 240 | 128 | 70 | 92 |
| 250 | 136 | 75 | 100 |

## STABILIZERS

Stabilizers are required in most of households because the voltage level is not maintained by the distribution companies. The variation in voltage level range from 140 to 250 V. This necessitates the use of stabilizers either at the mains or individually for refrigerators, TVs and other gadgets. The main component of the stabilizers is the step up/step down transformer. Although a transformer is treated to be an efficient device, the efficiency decreases as the quality of wires and the core material used are of inferior quality. Since these stabilizers are always connected to the mains, the losses are continuous in nature.

Table 5.4 shows comparison of power consumed by stabilizers of 1 kVA.

TABLE 5.4: POWER CONSUMED BY 1 KVA STABILIZERS

| Voltage (Volts) | At No Load | | | | At Load of 100 Watt Lamp | | | |
|---|---|---|---|---|---|---|---|---|
| | A | | B | | A | | B | |
| | Current (Amps) | Power (Watts) | Current (Amps) | Power (Watts) | Current (Amps) | Power (Watts) | Current (Amps) | Power (Watts) |
| 150 | 0.1 | 8 | 0.04 | 8 | 0.64 | 98 | 0.64 | 98 |
| 160 | 0.1 | 8 | 0.06 | 8.5 | 0.65 | 104 | 0.66 | 106 |
| 170 | 0.1 | 8 | 0.06 | 8.5 | 0.54 | 96 | 0.55 | 96 |
| 180 | 0.1 | 8.5 | 0.1 | 12 | 0.56 | 100 | 0.58 | 104 |
| 190 | 0.1 | 8.5 | 0.11 | 12.5 | 0.58 | 110 | 0.6 | 112 |
| 200 | 0.1 | 10 | 0.12 | 14 | 0.46 | 96 | 0.48 | 96 |
| 210 | 0.1 | 10 | 0.08 | 12 | 0.47 | 100 | 0.49 | 102 |
| 220 | 0.1 | 10 | 0.09 | 12 | 0.52 | 108 | 0.51 | 110 |
| 230 | 0.1 | 10 | 0.09 | 12.5 | 0.54 | 114 | 0.52 | 118 |
| 240 | 0.1 | 10 | 0.1 | 14 | 0.44 | 98 | 0.44 | 102 |
| 250 | 0.1 | 10.5 | 0.11 | 16 | 0.45 | 104 | 0.45 | 108 |

## INVERTERS AND UPS

A survey conducted in a residential area comprising about 1,000 houses found that more than 50 per cent of the customers were using inverters rated at 0.5 kVA. The inverter consumes a noticeable amount of power during charging and causes continuous power loss due to transformers installed in it. They are also a major source of generation of harmonics in the system and degrade the quality of power supply. Table 5.5 (*a* and *b*) shows the performance of different inverters and UPS.

TABLE 5.5a: PERFORMANCE OF INVERTERS

| Voltage (Volts) | At No Load | | | | Load (100 Watts) | | | |
|---|---|---|---|---|---|---|---|---|
| | Current (Amps) | | Power (Watts) | | Current (Amps) | | Power (Watts) | |
| | A | B | A | B | A | B | A | B |
| 150 | 0.40 | 0.55 | 40 | 50 | 0.92 | 0.80 | 30 | 104 |
| 160 | 0.60 | 0.62 | 68 | 54 | 1.13 | 0.88 | 45 | 112 |
| 170 | 0.80 | 0.68 | 104 | 56 | 1.38 | 0.91 | 112 | 120 |
| 180 | 0.85 | 0.71 | 108 | 59 | 1.43 | 0.97 | 124 | 128 |
| 190 | 0.92 | 0.74 | 116 | 60 | 1.50 | 0.99 | 144 | 134 |
| 200 | 0.98 | 0.78 | 120 | 60 | 1.54 | 1.00 | 160 | 140 |
| 210 | 1.01 | 0.79 | 122 | 62 | 1.58 | 1.02 | 172 | 148 |
| 220 | 1.06 | 0.80 | 126 | 62 | 1.60 | 1.03 | 192 | 156 |
| 230 | 1.08 | 0.80 | 128 | 64 | 1.63 | 1.04 | 204 | 162 |
| 240 | 1.10 | 0.80 | 132 | 64 | 1.65 | 1.04 | 220 | 170 |
| 250 | 1.14 | 0.80 | 130 | 64 | 1.66 | 1.05 | 232 | 176 |

TABLE 5.5b: PERFORMANCE OF UPS

| Voltage (Volts) | At No Load | | Load (100 Watts) | |
|---|---|---|---|---|
| | Current (Amps) | Power (Watts) | Current (Amps) | Power (Watts) |
| 150 | 0.25 | 10 | 0.75 | 104 |
| 160 | 0.36 | 10 | 0.81 | 114 |
| 170 | 0.48 | 10 | 0.89 | 126 |
| 180 | 0.24 | 10 | 0.66 | 108 |
| 190 | 0.31 | 10 | 0.70 | 116 |
| 200 | 0.41 | 10 | 0.76 | 124 |
| 210 | 0.20 | 9 | 0.58 | 106 |
| 220 | 0.21 | 9 | 0.61 | 116 |
| 230 | 0.30 | 10 | 0.65 | 124 |
| 240 | 0.38 | 10 | 0.50 | 108 |
| 250 | 0.20 | 10 | 0.58 | 120 |

### WATER PUMPS

The water level is going deeper day by day and municipal corporations are not supplying river water in many of the states. Households, therefore, have to make their own arrangements for water. To draw water from a deep borewell submersible pumps are most suitable. These pumps normally operate at a lagging power factor of 0.62 to 0.72. The control panel provided with low rating pumps does not have capacitors for power factor improvements. The customers have to install these capacitors.

The electromagnetic radiations produced by some energy conservation devices have not been studied so far. These electromagnetic radiations affect the health of users. We must provide some shielding and place these equipments at places so that radiations do not affect users. Extensive research is being in this are and results are expected soon.

We have estimated the money wasted in unwanted loss of power due to inferior quality electrical gadgets. For a ceiling fan of higher power rating and an efficient fan, if run 15 hours per day and the difference in power rating is 48 W, then (36 × 15 = 540 Watt hour). The energy wasted for say 8 months of fan use in a year will be roughly 122 kWh (i.e. Rs. 366 per year). This shows that if we spend more money on a good quality fan initially we can recover more than that money in the long run.

For a voltage stabilizer, the equipment is connected at no load for 24 hours and voltage input is 200 V, the total power loss in a day is 456 W or 167 kWh of energy per year, i.e. a loss of Rs. 525 (approx) per customer. This amount adds up to crores of rupees if all the stabilizers used in the country is taken into account. This loss will change further if we take into account the transformation loss at full load. The figure only refers to a 1 kVA stabilizer, but in actual practice the range varies from 0.25 kVA to 5 kVA more frequently. So if the power is available at constant voltage this revenue loss can be eliminated.

## CONCLUSION

Customers must be encouraged to use energy saving and energy efficient equipments. On the other hand, the regulatory commissions must ensure that the electricity tariff does not have a minimum unit charge to ensure energy conservation. The government must also ensure that electrical gadgets available in the market are of good quality.

The use of efficient lights, fans and stabilizers not only saves energy but also saves revenue. The energy thus saved can be used elsewhere. Further, if we maintain our distribution system, continuity of supply, and keep the voltage level at acceptable limits by installing suitable voltage compensators, the overall power crisis can be eliminated as well as the need for voltage stabilizers and UPS.

# 6

# *Application of Information Technology to Improve Performance in Power Distribution*

SANJAY GUPTA

Power distribution is the most critical segment of the electricity value chain. Given the poor performance of State Electricity Boards (SEB) in this segment, it becomes evident that the reform in the distribution segment is the key to the growth of an efficient and self-sustainable power sector in India. So far, several steps have been taken to improve the performance in distribution. Some benefits have been realized from these steps, however, these success have been limited in nature and changes have not been institutionalized to be sustainable over the longer period. Therefore, enabling infrastructure and proper systems are required to bring about sustainable and pervasive improvements. This requires bringing radical improvements across the entire distribution operations through improved operational processes leveraging best of information technology (IT).

Based on efficiency and customer satisfaction related issues in distribution, there are enough opportunities for application of IT in Indian power distribution. There is much to learn from the best practices in IT management and IT applications from around the globe. At the same time, we have to develop a holistic long-term approach for IT implementation suited for Indian conditions. This approach should include prioritization of opportunities based on business cases to ensure that each initiative to be undertaken is applicable specific to the Indian context.

## DISTRIBUTION: THE MOST IMPORTANT SEGMENT OF VALUE CHAIN

The financial and operating performance of SEBs is an issue of concern for everyone. The financial losses[1] of SEBs have reached to the tune of Rs. 26,000 crore per year primarily due to power theft and the inability to bill and collect. Losses due to power theft alone are approximately Rs. 20,000 crore per annum. Numerous published data and reports indicate transmission and distribution (T&D) losses to be in the vicinity of 25 per cent; however, indirect estimates and recent studies by various states put these losses at a much higher level of around 40-50 per cent. By comparison, T&D losses in the developed countries are typically 7-8 per cent. Even in developing countries such as China and Thailand, T&D losses are at a much lower 8-12 per cent range.

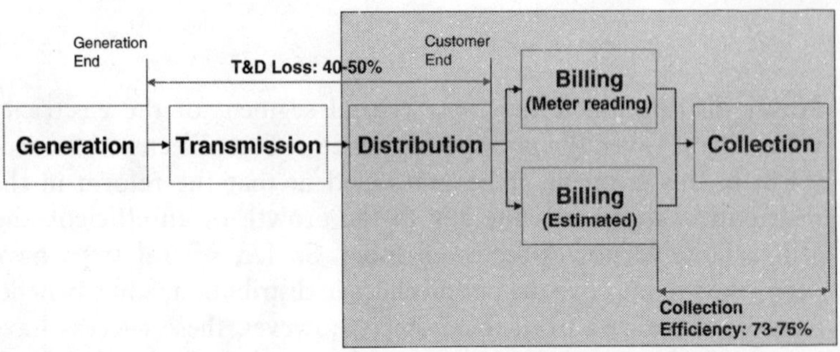


FIGURE 6.1: TYPICAL VALUE CHAIN FOR A SEB

Figure 6.1 depicts the typical value chain for a SEB in India. The activities covered under the shaded part responsibilities of the distribution segment of SEBs. The criticality of distribution in the electricity value chain is amply emphasized by the following facts related to operational efficiency and customer satisfaction:

- At the national level, only 55 per cent (Rs. 62,000 crore) of the total energy generated is billed out and the rest is classified as T&D losses. Around 80-90 per cent of these T&D losses take place in the distribution segment. Around two-thirds of the total T&D loss in distribution is commercial loss, which takes place mostly at 11 kV and below—distribution has operational control at 33 kV and below.

- Commercial losses are due to rampant theft and pilferage of electricity, meter tampering, unauthorized connections and unmetered supply—avoidance of these activities falls under the preview of distribution.
- Only 74 per cent of billed energy (Rs. 46,000 crore) is realized by SEBs. Billing customers for the electricity consumed and collection of revenue from customers are activities also associated with distribution segment of the value chain.
- Reliability and quality of supply to customers is very poor with frequent interruptions and wide fluctuations in voltage and frequency.
- Customer services are substandard due to lack of proper systems for handling customer grievances such as outages and billing inquiries.

Due to these factors, stakeholders are not satisfied. Private investors are wary of investing and the public sector generation companies have their funds tied up in huge outstanding payments with the SEBs. Customers are not happy with the supply position and increasing tariffs due to SEBs' inefficiencies.

The above makes it evident that poor performance of SEBs in distribution is hampering the growth of an efficient and vibrant power sector in India. Improved performance in distribution is more crucial because inefficiencies in this segment mean higher tariff and unreliable supply to customers.

## DISTRIBUTION REFORMS: ROLE OF IT AS ENABLER

Reform in the distribution sector should be the key focus area to realize the Ministry of Power's (MoP) vision of reliable, affordable and quality power for all by 2012. Several steps have been taken by the MoP and various states under the guidance of MoP in this regard. Some of the steps taken are:

- Unbundling of integrated entities into generation, transmission and distribution and corporatization of these entities;
- Appointment of regulatory authorities;
- Division of SEB distribution into several manageable companies and then privatization of these entities in a few states;
- Specific drives to enhance revenue and reduce losses;

- Substantial investments to enhance the quality of networks and drives to installs meters;
- Distribution circles as centres of excellence under Accelerated Power Development and Reform Programme (APDRP).[2]

These steps have brought about some benefits such as improved operating structures, increased commercial orientation, increasing transparency in operations and increased customer orientation. However, the impact of these has been limited to a few pockets and there has been limited success in institutionalizing these changes for sustainability. As a result, the distribution segment continues to be plagued with issues such as revenue leakages, irregular supply and customer dissatisfaction.

Therefore, the need is to create enabling infrastructure and systems to bring about sustainable and pervasive improvements. This requires bringing radical improvements across the entire distribution operations through improved operational processes leveraging best of breed IT. Execution of core business processes at the transaction level using information systems should lay the foundation for use of IT. This would also ensure better, faster and accurate decision-making at all levels in the organization as shown in Fig. 6.2.

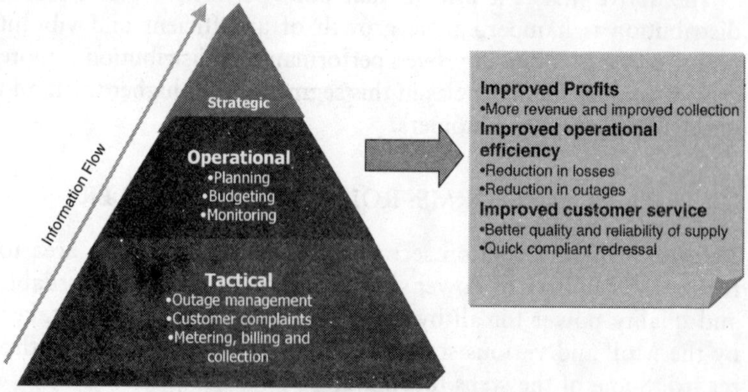


FIGURE 6.2: INFORMATION TECHNOLOGY AS TRANSFORMATION AGENT

IT can be of immense value in transforming the distribution sector by improving efficiency, increasing productivity and reducing losses as shown in Fig. 6.2. Some of the benefits of leveraging IT in distribution are:

1. Profitability improvement through
   - Revenue enhancement;
   - Cost reduction;
   - Commercial and technical loss reduction;
   - Shortened consumption to collection time.
2. Enhanced customer satisfaction
   - Improvement in reliability and quality of service;
   - Improved handling of customer complaints.
3. Other gains
   - Operational efficiency through effective decision support;
   - Flattening of peak demand curve by better load management.

## DISTRIBUTION: PROCESSES AND ISSUES

The key functional areas and processes in electricity distribution are shown in Fig. 6.3. How can we leverage IT to improve performance in distribution? For this, we need to analyse these processes and identify issues pertaining to these processes. Issues associated with key processes for SEBs are identified below.

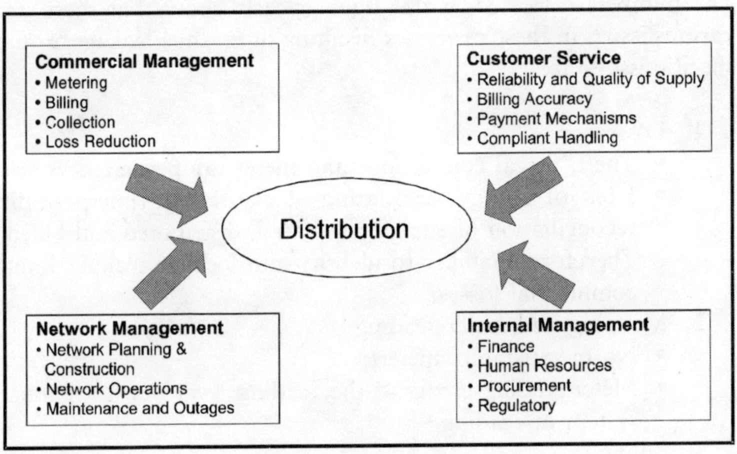


FIGURE 6.3: FUNCTIONAL AREAS AND PROCESSES IN DISTRIBUTION

### COMMERCIAL MANAGEMENT

Metering, billing, collection and commercial loss reduction are key processes in commercial management. Fig. 6.4 shows the

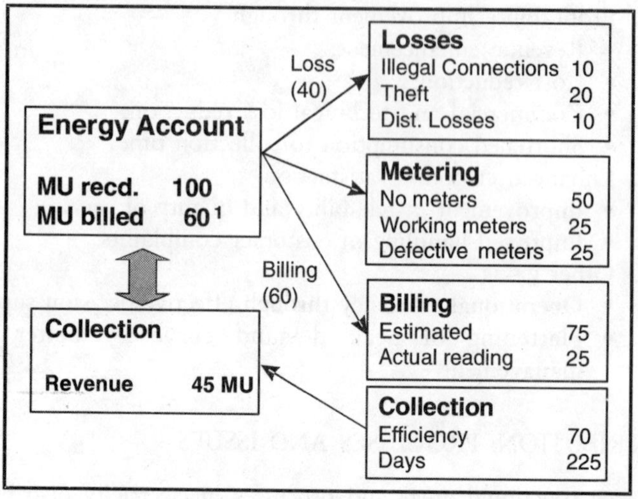


FIGURE 6.4: COMMERCIAL MANAGEMENT FOR A TYPICAL SEB DISTRIBUTION

performance of a typical SEB distribution entity on these processes. The numbers indicated in this figure clearly show that there are various issues in these processes resulting in revenue leakage. Some specific issues are:

1. Losses
   - Theft, illegal connections and meter tampering;
   - Lack of energy accounting—lack of real-time/periodic reconciliation of energy purchased, distributed and billed. Therefore, inability to identify and localize technical and commercial losses.
2. Metering and meter reading
   - No meters/faulty meters;
   - Meter-reading errors at the reading level due to manual system of reading.
3. Billing
   - Invoicing errors and higher percentage of estimated billing;
   - No billing for new customers.
4. Unpaid energy (Collection losses)
   - Lack of monitoring and analysis.

CUSTOMER SERVICE

Fig. 6.5 shows four key areas of concern for customer service. Improvement of performance in these areas will improve customer service levels. Specific issues associated with these areas are:

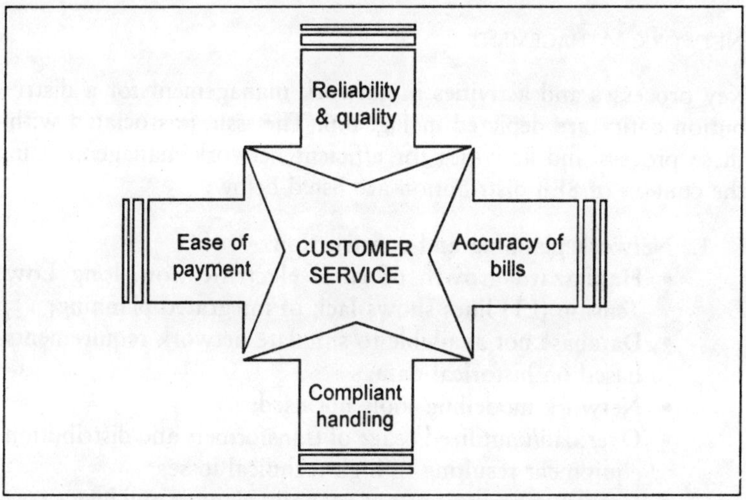


FIGURE 6.5: AREAS OF CONCERN IN CUSTOMER SERVICE

1. Quality and reliability of supply
   - Large number of scheduled and unscheduled outages;
   - Lack of communication from SEBs on possible outages and outage time;
   - Response to trouble call non-customer friendly.
2. Accuracy of bills
   - Inaccurate meter reading or no meter reading or reading taken but not reported;
   - Changed meter status or split between cycles;
   - Wrong calculation;
   - No posting of collection.
3. Payment mechanisms
   - Bill not delivered in time to make payment;
   - Waiting time for bill payment;
   - Preferred vs. available mode of payments.

4. Compliant handling
   - Demand vs. provision of new connections;
   - Responsiveness to customer queries and specifically to bill queries;
   - Preferred vs. available facilities to log complaints.

## Network Management

Key processes and activities in network management for a distribution entity are depicted in Fig. 6.6. The issues associated with these process and activities for efficient network management in the context of SEB distribution are listed below:

1. Network planning and construction
   - Haphazard growth of rural electrification, long Low Tension (LT) lines shows lack of integrated planning;
   - Database not available to simulate network requirements based on historical data;
   - Network modelling tools not used;
   - Overload/unutilized usage of transformers and distribution equipment resulting in high technical losses;
   - Construction projects not properly planned and hence cost and time overrun.
2. Network Operations
   - Real-time network operating parameters not available at distribution level to take corrective action;
   - Lack of automation, feeder levels 66 kV onwards till substations;

| Network Planning & Construction | Network Operations | Network Maintenance |
|---|---|---|
| • Network planning<br>• Design<br>• Work management<br>• Project management | • Control & automation<br>• Alarm & event handling<br>• Switching operations<br>• Load shedding | • Asset management<br>• Crew management<br>• Trouble call handling<br>• Outage analysis |

FIGURE 6.6: KEY PROCESSES AND ACTIVITIES IN NETWORK MANAGEMENT

- Database not available, therefore prediction of demand vs. time for specific areas not possible;
- Varying load demands result in load shedding and disruptions, also damaging equipment.

3. Network Maintenance
   - Database for asset inventory and their preventive maintenance schedule not available;
   - Equipment location, circuit/feeder diagrams not easily available for maintenance;
   - Database not available to analyse outages/equipments' failure;
   - Preventive maintenance policy not in practice;
   - Allocation of resources—manpower and material—is not optimal.

INTERNAL MANAGEMENT

1. Finance
   - Manual compilation of accounts;
   - Data inconsistency and errors;
   - Lack of timely and accurate information for decision-making.
2. Human Resources
   - Lack of proper employee records, benefits and skills;
   - Manual records of employee history.
3. Procurement
   - Manual and time taking procurement procedures;
   - Vendor selection process and vendor management issues;
   - Inventory positions not easily available.

## INFORMATION TECHNOLOGY TO IMPROVE PERFORMANCE IN DISTRIBUTION

Having discussed the key issues and the key elements to be addressed to improve performance in distribution, we need to identify IT applications that can help SEBs to move in this direction. The relevant IT applications in various functional areas of distribution are listed below:

COMMERCIAL MANAGEMENT

| | Process Requirements | IT Applications |
|---|---|---|
| Metering and meter reading | • Develop an inventory of meters, locations<br>• Prepare a meter reading route plan with optimal allocation of readings in a route and cycle<br>• Download customer data on MRI for reading | • Install electronic meters with optical port for direct download facility through Meter Reading Integration (MRI)<br>• Meter inventory system<br>• Meter reading system<br>• MRI for noting all kind of readings |
| Billing & Collection | • Collect meter reading information from meter reading system<br>• Use tariff from customer information data base<br>• Automate process of billing for all customers<br>• Send revenue data to collection system<br>• Collection analysis and send outstanding information to field | • Billing application<br>• Integrate with meter reading system<br>• Collection system<br>• Integrate collection system with billing system<br>• Integrate collection system with field collection units |
| Energy Accounting | • Metering of all sub-stations up to 11 kV including Distribution Transference (DT) primary in first stage<br>• Metering on DT secondary at 0.4 kV level<br>• Calculation of energy losses at feeder level in the beginning and then at DT level based on billed revenue<br>• Mapping of customers to feeders and DTs<br>• Energy loss analysis | • Install electronic meters with remote reading facility at all sub-station feeders, interconnections<br>• Install electronic meters with remote reading facility at DTs in later stage<br>• Energy accounting system<br>• Integrate energy accounting system with billing and customer information database<br>• Automatic Meter Reading (AMR) for all customers |

NETWORK MANAGEMENT

| | Process Requirements | IT Applications |
|---|---|---|
| Network Planning and Construction | • Collection of current network data, equipment details<br>• Load growth analysis<br>• Scheduled system maintenance to be informed to customers | • Network planning and analysis tools<br>• Forecasting tools<br>• Network simulation software<br>• Integrate design with simula- |

| | | |
|---|---|---|
| | • Automate the designing process<br>• Network simulation<br>• Improve project management through project structure and monitoring system | tion to analyse network changes<br>• Work management system<br>• Project management system |
| Operations | • Real-time network data, alarms and alerts<br>• Automatic switching operations<br>• Location and equipment information<br>• Outage analysis | • Start with SCADA and then progressively move to sub-station automation and then feeder automation<br>• Interface to sub-station control and protection system and then to feeder automation system<br>• Integrate SCADA with Geographical Information System (GIS) and asset system to identify asset properties and location<br>• Integrate with trouble call and Customer Information System (CIS) to complete distribution management system |
| Maintenance and Outage Handling | • Call recording for outage<br>• Centralized crew management and job allocation<br>• Enable crew with mobile devices and vehicles<br>• Outage analysis<br>• Develop database of all the asset and their predetermined maintenance schedule | • Trouble call management system<br>• Integration with emergency call handling resources<br>• Field force automation by integrating field mobile devices with control centre<br>• Asset management system<br>• GIS<br>• Outage management system |

CUSTOMER SERVICE

| | Process Requirements | IT Applications |
|---|---|---|
| Quality and Reliability of Service | • Need of real-time system data for proper system maintenance and faster action on outage<br>• Identification of customers affected by outage and information to customers regarding outage and possible duration of outage | • SCADA and distribution automation<br>• Outage and work management system including trouble call management system for faster restoration of outages<br>• GIS for identifying area and equipments<br>• Customer information thr- |

| | | |
|---|---|---|
| | • Scheduled system maintenance to be informed to customers | ough internet, phones, call-centres<br>• IVR for complaints and call back on restoration of supply |
| Accuracy of Bills | • Better designing of meter reading process<br>• Reduce manual intervention in meter reading process<br>• Automation of meter reading and billing process | • Download devices for meter readings<br>• Meter reading interface with billing applications<br>• Billing applications |
| Ease of Payment | • Develop internet-based payment system<br>• Develop innovative payment options such as budget payment<br>• Multiple payment locations and not restricted to a particular division or sub-division<br>• Implement proper credit process and develop credit control & monitoring process and system | • Electronic payments, including direct debit payments<br>• Prepaid metering<br>• Bar coding facility even when payment is made in person, for faster payment<br>• Develop proper credit control and collection system<br>• Integrate banks in utility payment process |
| Compliant Handling | • Develop contact centres (call centres) with customer database and related information<br>• Develop new connection process and automated process | • Call centres with IVR for outage as well as bill-related complaints<br>• Call back from IVR on redressal of any bill related complaint |

### Internal Management

For improving internal management, a transaction system would be required so that once a data is entered, the same data can be used for all subsequent purposes. Implementation of Enterprise Resource Planning (ERP) system will help in reducing all issues related with internal management areas such as finance, HR and procurement.

### Integration

The list of applications mentioned shows a lot of commonality across functional areas and a need for integration. In fact, these functions are not isolated and for efficient distribution operations, data flow is required from one function to another function. For example, a well integrated CIS can handle most of the commercial and customer service requirements (Fig. 6.7).

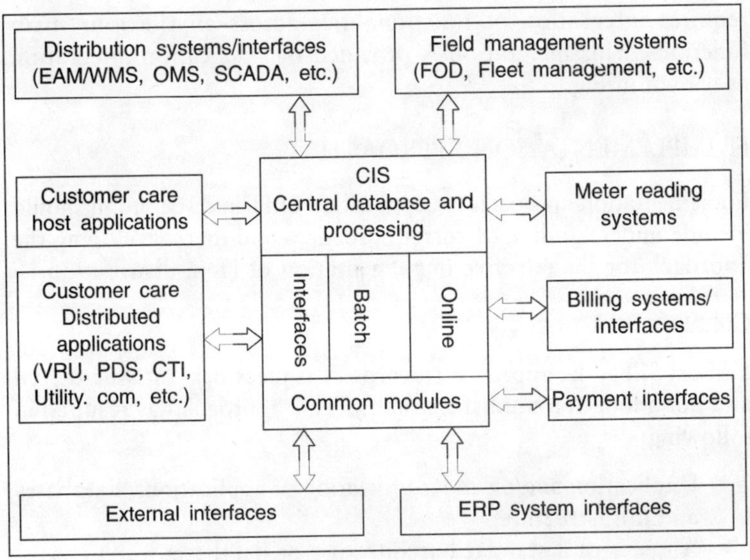


FIGURE 6.7: CUSTOMER INFORMATION SYSTEM FOR DISTRIBUTION OPERATIONS

Similarly, an efficient and effective network management for improved reliability and customer satisfaction and to reduce cost

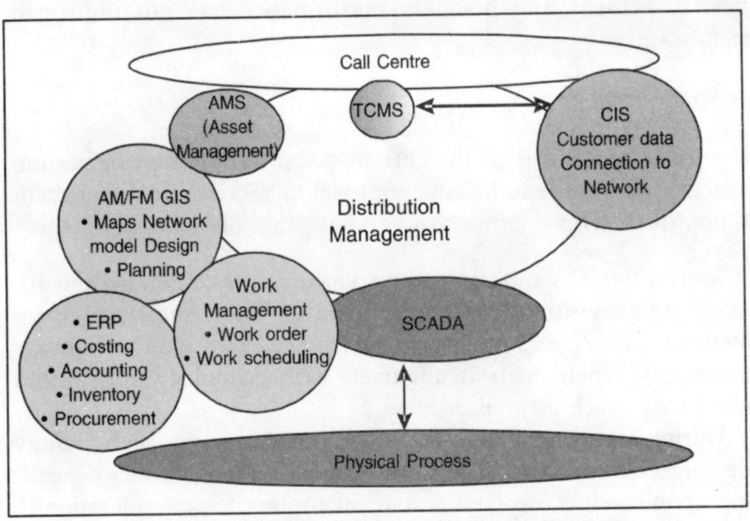


FIGURE 6.8: INTEGRATED NETWORK MANAGEMENT (INPUTS FROM ABB)

requires integration of functionalities across all the four main functions. This integration is provided by application integration as shown in Fig. 6.8.

## IT IMPLEMENTATION APPROACH

Implementation approach for application of IT in distribution should include understanding of current practices and then developing the approach for the effective implementation of IT in distribution.

### CURRENT PRACTICES

In most SEBs, the approach towards IT is piecemeal or issue-driven and not a long-term holistic strategy. This approach has resulted in following:

- Duplication and/or under-utilization of applications, databases and infrastructure;
- Absence of a standard architecture or database;
- Systems in place are largely aggregators or processing tools with a lot of manual inputs and without required controls;
- Limited integration of systems, multiple sources.

Due to above-mentioned issues, IT investment has not provided desired returns and business performance has not improved significantly.

### IMPLEMENTATION APPROACH

Due to shortcomings of the current piecemeal IT implementation practices, a long-term holistic approach is required for implementation of IT. This approach should have the following key steps:

*Prioritization:* Organizations should have prioritization framework in place to prioritize IT opportunities after 'as is' and requirement analysis. This should be based on several factors such as business needs, cost benefit analysis, alignment with technology strategy and risk factors such as preparedness.

During cost-benefit analysis, we also need to decide on build/buy decision. This decision needs to be made broadly as a policy depending upon product availability and time-to-market considerations.

Prioritization could also be for various time periods and for any organization this can be divided into the following four categories:

- Prerequisite—necessary first steps to be established first to provide for implementation of subsequent IT applications such as:
    - Development of customer information database;
    - Database for meter inventory;
    - SCADA;
    - Asset inventory and network mapping database.
- Short-term—quick wins such as commercial applications
    - Billing application;
    - Credit and collection;
    - Finance and accounting;
    - Integrated metering, billing and collection systems with downloadable meter reading;
    - Energy accounting system to conduct energy audits.
- Medium-term—cover core processes such
    - Call centre;
    - Service connection and maintenance processes;
    - Asset and work management, outage management;
    - Material management.
- Long-term—Advanced applications such as AMR and distribution management could be built in this phase.

*IT Implementation Plan:* Emphasis should be on adopting a holistic approach towards the standardization of the following IT areas:

- Architecture
- Applications
- Network
- Hardware.

*Organizational Plan:* Organizational plan to support IT implementation should have:

*Realigning Responsibilities and Authority:* IT implementation may result in centralization of certain tasks and activities such as billing, which can be centralized at higher levels. It will require change in responsibilities and authority and hence change in power equations within the organization.

*Change Management:* Large-scale IT implementations result in changes in the way the employees work, the way they are measured and deliver services to customers. This requires a fundamental change in mindset for the IT implementation to succeed. A well-designed and well-implemented system can only provide data and information to managers for decision-making but in the end the decisions will have to be taken by the managers using the data.

*Sustained Leadership Commitment:* Commitment from top leadership of the organization has to be shown towards importance of IT for the enhanced performance of the organization. This should also result in the creation of best in breed processes and systems and supported by best of the IT and managerial talent.

## CONCLUSION

Application of IT in distribution utility has enormous potential to transform the business and improve customer experience given the small use of IT by distribution organizations in India compared to global distribution utilities. In global utilities, IT plays a central for execution of business processes as well as decision-making at every level in the organization.

Investment requirements for application of IT in distribution utilities to attain global standard can be huge. However, given the precarious financial position, utilities cannot make this much investment in one go. Hence, utilities in India need to adopt a phased approach towards IT implementation and this investment in IT should be purely driven by business needs and quick returns. This will also require business-IT integration and a holistic long-term strategy for the adoption of IT.

## NOTES

1. Financial parameters at the end of 2000-1.
2. Blueprint for Power Sector Development, MoP (2001).

# 7

# *Broad Technology Choices and Technology Mix for the Power Sector*

R.K. BELAPURKAR

## BACKGROUND

The technologies for providing clean and quality power at affordable as well as remunerative levels have to be evolved keeping in view the efficiencies of fuel use and ever increasing environmental concerns. The power equipment manufacturing industry has been trying to cope with these challenges for over a century now. Rapid industrialization during the twentieth century led to a spurt in demand for power all over the world. The power plant equipment manufacturing industry made technological advances to keep pace with this requirement by utilizing a variety of fuels.

With the building up of 'spinning reserves' in the developed countries over the years, the market for power generation equipment has shifted from the developed world to the developing countries to meet the growing demand for electric power in these regions.

## INTRODUCTION

The principal uncertainty in global energy supply prospects is cost. Advances in technology and productivity are driving production and transportation costs lower, but the depletion of the cheapest reserves and the growing distances over which new supplies must be transported are in many cases, pushing delivered energy costs up. The other factor in the energy supply picture is the price. Energy prices play a major role in determining the timing and amount of investment that goes into expanding energy supply capacity. Financing for the development of energy infrastructure is a major challenge. Massive investment in the production, transformation, transportation and distribution of energy will be needed to meet the growing demand. The bulk of this investment is needed in the developing countries. Mobilizing this investment in a timely fashion will require the lowering of regulatory and market barriers.

Global energy supply trends have major implications for the governments of producer and consumer countries alike. Addressing supply—security issues by creating appropriate regulatory and market frameworks and in encouraging technology development and deployment shall confront the governments. Harmonization of trade and tariff rules will be especially important to cross-border pipeline projects. Research and development will be vital to reducing energy supply costs.

The most uncertain factor affecting future fuel supply is the impact of environmental policies on demand. Technology is expected to drive continuing improvements in efficiency and reductions in the cost of fossil fuel extraction and preparation. The extent to which governments encourage technologies that generate low or zero carbon emissions and the costs involved are key issues in the long term. How much this will cost is uncertain.

## GENERATION SCENARIO—DEVELOPING COUNTRIES

In the next 25 years, as much additional power generating capacity may be added around the world as was built in the entire past century, with China, India and the Pacific Rim Countries accounting for a major share of this addition. China's generating capacity may increase from 3,15,000 MW to more than 5,00,000 MW by 2010. India is planning 1,00,000 MW between 2002 and 2012. Malaysia is likely to add 10,000 MW in 10 years, Philippines 10,000 MW by 2009 and other South-East Asian countries like Taiwan, Thailand and Indonesia all planning substantial capacity additions.

Demand for electric power is an index of growth in a developing economy. Notwithstanding the fact that the total installed generating capacity in India has increased substantially, the demand still outstrips generation. While in the recent past the annual capacity additions have remained around 3,500 MW due to constraints on resources and investment flows, to make power available on demand (zero deficit) it is estimated by the Government of India to add 1,00,000 MW by the year 2012. India is one of the very few developing countries which produces a full range of electric power generation and transmission equipment.

Let us look at the future fuel option for power generation. Globally, the future fuel options will be based on three broad categories, viz., fossil fuels, nuclear energy and the renewable sources. Choice and mix of fuel will depend mostly on the local conditions of availability, cost of transportation, environmental considerations and costs of production. No credible energy strategy can realistically postulate that the world is going to change its energy mix dramatically for a considerable period into the twenty-first century. Coal is a very important component of our energy future and because it is, it will be a very important part of our forthcoming energy strategy. Coal is where our significant capacity addition is targeted. Natural gas will also play an important role with substantial increase in its use in the regions where the same is available as a cleaner means of energy, mainly for meeting peak load conditions. Globally, gas reserves are sufficient to meet the demand for next several decades in the long term; however, large-scale fuel switching from coal to gas does not seem optimistic even at the global scale. Oil is likely to continue as an energy source, specially for standby or peaking plants. With a major thrust propounded by the government, hydro power is expected to get an acceleration, particularly in the north-eastern region. Nuclear fuels like uranium will find application only in a limited way.

The resource base of the renewable energies, particularly solar and wind, is so immense that they could supply more energy than the world demand today once there are technologies in place which can harness these economically. The fuel mix in each region will drive the technology of the power plant equipment to be used for power generation.

As world energy consumption rises, advances in energy technologies are taking place in clean power generation and efficiency. Global trends in pollution-free and high efficient technologies indicate that the following technologies will dominate the generation scene in the twenty-first century in addition to conventional fossil fuel-based power plants, which can be broadly classified into three categories:

(*a*) Technologies substantially developed
- Super critical thermal power plants

- Coal-based combined cycle plants
  —Integrated Coal Gasification Combined Cycle (IGCC)
  —Pressurized Fluidized Bed Combustor (PFBC)
- Advanced gas turbines-based power plants
- Nuclear power generation with fast breeder technologies.

(*b*) Developing technologies
- Fuel cells
- Micro turbines
- Renewable energy sources like wind energy and solar photo voltaic.

(*c*) Futuristic technologies
- Fusion power
- Geo-thermal, wave/tidal

## Super Critical Thermal Power Plants

With increased temperatures and pressures, super critical boilers, besides having higher efficiency, also respond better to load changes, and in addition reduce emission of pollutant greenhouse gases like $CO_2$, $SO_x$, $NO_x$ and particulate matter.

Present-day efficiency of new power plants with high steam conditions even exceeds 43 per cent. The main thrust has been in the development of materials to withstand higher temperatures in the range of 540 °C to 650 °C. In the area of steam turbines, latest developments are towards use of improved blading, reduction of cylinders from three to two and even to single cylinder to reduce cycle times in manufacturing, erection and commissioning. However, cost of coal would drive this technology to its economical advantage.

## Coal-based Combined Cycle Plants

In India coal shall continue to stay as primary energy source during the twenty-first century, since the country is blessed with abundant coal reserves. The Indian coal has high ash content besides abrasive silica. Therefore, due to its obvious effects on the environment, to maintain low emissions of sulphur oxides and carbon-dioxide, better technologies with higher conversion efficiency are required. The international trend is towards advanced coal-fired power generating systems and clean coal technologies like:

- Integrated Coal Gasification Combined Cycle (IGCC) power generation for clean coal usage. Overall efficiency of an IGCC plant is about 3-5 per cent higher than the conventional Rankine cycle for identical capacity. For high ash Indian coals, with ash fusion temperature around 1,400 °C, fluidized bed gasification will be techno-economically more appropriate and can be expected to be the viable technology for twenty-first century. Based on this experience, BHEL has plans to set up a 100 MW coal-based IGCC demonstration plant with assistance from the Government of India. With improved efficiency, $CO_2$ which accounts for half of the total greenhouse emission can be directly reduced. BHEL has developed low $NO_x$ gas turbine burner which is already a commercial product. Since the incidence of sulphur in Indian coals is generally low, $SO_2$ emission will also be well within the environmental regulation.
- Pressurized Fluidized Bed Combustion (PFBC) power generation emphasizing both sulphur removal from furnace and improved efficiency. Efficiency levels of PFBC-based power plant will be around 2-3 per cent less than IGCC. Reduction of sulphur emissions during combustion can be achieved by adding limestone to the bed materials. The main technological challenge in PFBC technology is the cleaning of hot gases.

However, economics of power generation have still to mature for these technologies.

### Advanced Gas Turbines

Currently gas turbines of sizes up to 160 MW are in use in the country. However, advanced gas turbines have become a leading source of power generation and will make further strides in the coming years with increased ratings through improved metallurgy, protective coatings, effective cooling methods, significant improvements in efficiency to the order of nearly 60 per cent coupled with lower emissions.

These turbines are versatile in terms of using a wide array of fuels such as natural gas, coke oven gas, refinery gas, naphtha, distillate, kerosene, and so on.

### Nuclear Power Generation

The Nuclear Power Corporation (NPC) has a vision—to attain 20,000 MW by 2020. There is a need to evolve foolproof safety measures, particularly after the accidents in technologically advanced countries like the USA, Russia and more recently Japan.

### Renovation and Modernization

The power plants which were set up prior to the 1990s were with technologies in which rapid advancements have now taken place in terms of increased efficiencies and producing cleaner and quality power. It is now required to retrofit the existing power plants, after necessary studies, with user-friendly and accurate controls, energy efficient systems and pollution control devices. In addition, life extension of the older plants should also be undertaken where the costs are expected to be about one-third of a new plant.

Central and state utilities, besides the private sector would be compelled to look at this option to quickly benefit from the low cost capacity and efficiency improvement programme, which is likely to be unleashed from this source. With the thrust by the Government of India for funding R&M programmes at concessional interest rates, a number of renovation and life extension studies have been initiated and these are being taken up rapidly for implementation.

### Fuel Cells

Fuel cells are electrochemical energy converters and have long been recognized as having excellent features such as highly efficient energy conversion, low carbon-dioxide emission and potential application to both small- and large-scale plants. Of the many types of fuel cells mainly characterized according to the electrolyte use, Phosphoric Acid Fuel Cells (PAFC), Molten Carbonate Fuel Cells (MCFC), Solid Oxide Fuel Cells (SOFC) and Proton Exchange Membrane Fuel Cells (PEMFC) find application for power generation. Developments are rapidly taking place in the world, led by Japan and the USA.

In India, BHEL has been leading the R&D efforts in PAFC and has successfully developed and demonstrated 1 kW and 5 kW PAFC power modules in 1991 and 1995 respectively at a chlor-alkali industry. Currently, with the Ministry for Natural Energy Sources (MNES) support, 50 kW PAFC power module is in an advanced stage of development and performance evaluation is in progress for a 200 kW PAFC power plant. It is expected that BHEL will be in a position to design and supply higher capacity fuel cells in the near future. This technology is likely to find commercialization in the next few years provided it is adequately supported by the respective governments of the world.

### MICRO TURBINES

Distributed generation (DG) technologies have the greatest potential in areas where there is no local utility or where power generation is unreliable or highly priced. Mainly because of lower capital cost, micro turbines hold the most promise of all DG technologies for application today with very high growth projections. Micro turbines are approaching the threshold of economic viability and efficiencies approaching 30 per cent. Even though the capital cost for Micro turbines is higher, it can be made up in more reliable performance and lower maintenance costs over the unit's lifetime.

### RENEWABLE ENERGY SOURCES

The clean energy sources such as solar, wind, tidal wave, ocean thermal, geothermal, hydrogen and are almost unlimited in amount, besides being environmentally friendly. However, excepting solar and wind power technology, many of these options are still in an experimental stage.

#### WIND ENERGY

The thrust on wind energy in several countries like the USA, Germany, UK, Sweden, Canada and India can be traced to the oil crisis in the early 1970s. These have been largely driven by tax credits and renewable energy targets. In India, these machines were introduced in 1986. Presently, the largest machine in India is rated 600 kW while internationally, 1.2 MW and 1.5 MW machines are being introduced.

India has the third largest installed capacity in the world after the USA and West Germany. As per MNES estimates, the capital cost of a wind power station is around Rs. 4.0 crore per MW and thus compares favourably with thermal power generation. India also has a vast wind energy potential, estimated to be around 20,000 MW.

### SOLAR PHOTO VOLTAIC POWER

Photo voltaic (PV) technology is well established as a reliable source of electricity in small, scattered applications like telecommunications, data acquisition, microwave repeater stations, space power and medium rating grid-connectable systems. Worldwide sale of PV systems reached 150 MW in 1998, after a decade of growth at 15-30 per cent per year. Extrapolating to 2023—half a century from the 1973 oil crisis and postulating a steady growth of 25 per cent—PV is expected to add a total of 40,000 MW of new generating capacity by that year. PV industry in the United States anticipates that about 15 per cent of the added capacity needed each year in the early 2020s will be covered by PVs.

Where governments are subsidizing the installation of PV arrays, growth may be even faster. Japan has maintained a large and steady R&D effort for decades and recently begun an initiative to install 5,000 MW of PV by 2010.

BHEL with an established annual capacity of 2 MW has supplied PV systems for rural radio telephones, street lighting, water pumping, railway distant signalling, microwave repeater stations, gas detection and telemetry on offshore platforms, based on mono-crystalline solar cell technology.

Presently PV systems are not cost-effective and cannot compete with the conventional power generation sources. However, developments in this field like gallium arsenide solar cells, and thin film technologies are expected to reduce the costs and PV systems can find wider applications. This is one of the renewable energy sources which has the greatest possibilities to become an alternate source of energy to conventional sources, in the very near future.

## FUTURISTIC TECHNOLOGIES

### FUSION POWER

The quest for attaining controlled fusion power in the laboratory has been described as the most challenging scientific endeavour and the ultimate solution to the growing energy needs of the world. Fusion energy has the advantages of inexpensive and practically inexhaustible fuel (deuterium occurs in water and tritium can be processed from lithium), besides the absence of radioactive by-products. The Indian fusion reactor, called Aditya, based on Tokomak technology, was commissioned in September 1989 at the Institute of Plasma Research, Gandhinagar. However, several technological issues need be resolved before fusion power can become commercially viable. The requirement of financial resources has been put at around US $50 billion in 20 years. However, its promising potential, and environment friendliness make this a technology worth pursuing.

## TRANSMISSION AND DISTRIBUTION

An age old cliché 'Energy saved is energy generated' aptly fits into the regimen of efficient power transmission. The planning of networks, i.e. establishment of new lines, voltage levels, routing and location are extremely important but are also guided by socio-economic considerations. One of the major equipment in a transmission line is the transformer, which works with an efficiency of over 99 per cent.

With the use of Static VAR Compensation systems (SVC), the required amount of reactive power is generated under varying load and system conditions. It also helps improve voltage, thereby further reducing losses. Series capacitors reduce transmission line impedances, thereby allowing increased power flow in the lines. Thyristor Controlled Series Compensation (TCSC) provides control over the power flow, helping in better routing of power flow in the interconnected grid, lowering the losses further.

A recent development is the Controlled Shunt Reactor (CSR) which replaces fixed shunt reactors employed in HV systems. A CSR remains in circuit only for the required duration, while still being able to come back into circuit as soon as required. BHEL has installed a 50 MVAR CSR in a sub-station of PGCIL at Itarsi and the feedback is excellent. Phase Shifting Transformers (PST), STATCOM (static compensator), a further refinement of SVC scheme, are also available to have better control of power flow. PST vary the phase angle between two points in the line, thereby affecting the power flow on that line. STATCOM is a static SVC which eliminates physical capacitors and reactors (as required in SVC) and helps in better control characteristics and reduces losses.

All the above schemes are part of Flexible AC Transmission System (FACTS), which has been made possible with the advent of power electronics and microprocessor-based control systems. Sharing of power between surplus and deficit regions with differing time zones and different peak load times effectively utilise the commercially available interconnection option of HVDC.

765 kV transmission system has been on the anvil for some time now. Two lines have been constructed but charged at only 400 kV. Another line is also under construction but will be charged at 400 kV. A new 765 kV line is being taken up, which should further improve the transmission system.

However, following technologies are being widely considered as future transmission options in the world:

- Superconducting Magnetic Energy Storage (SMES)—Size and expense are considered limiting factors. However, it is considered most appropriate and attractive for grid support applications.
- Ultra Capacitors (UC)—This technology is in the laboratory research stage. Developments are promising and is likely to compete with established and mature technologies.

For the near future, the Electricity Act 2003 allows a number of consumers to come together to form cooperatives for own use and sell any surplus power through a trading mechanism. Such stations could be established at pit-head or anywhere in the country, with the power being wheeled to the consumers through the interstate and intrastate grids or through dedicated transmission lines. This would provide incentive for setting up small captives thereby driving away the best paying industrial consumers from the SEBs, fold due to cross-subsidization.

## CONCLUSION

The R&D spending has to be stepped up considerably to focus on critical new energy technologies, accelerating investments in the new devices and channelling international energy assistance to developing countries. Globally, the governments will need to set the rules, focussing on ways to ensure that environmental costs are considered when economic decisions are made.

A major global concern today is the need to generate clean and green energy. We shall have to take immediate steps by inducting modern, highly efficient, pollution-free equipment and systems in all the existing as well as new plants.

Presently the thermal/hydro mix ratio in India is 75 to 25 and hydro power meets the peak load requirements of the grid. India has a vast hydro potential. Due to the high initial capital costs as well as problems associated with obtaining environmental clearances and long gestation periods, hydro projects do not receive the required priority. However, the associated benefits of low O&M costs, irrigation advantages and green power must get the due weightage and a better thermal/hydro mix is an essential objective which must be achieved.

India's requirement for power is immense, needing massive investments. The climate of industrial liberalization and the size of the potential market has resulted in the entry of a number of global players. The expectation of customers is also changing. In this scenario, it is essential that the Indian power industry becomes proactive and takes strategic initiatives to develop and offer the best deal to customers. Fortunately, BHEL's experience with equipment that can be designed to the varying requirements of Indian coal and different conditions of grid operation and experience of servicing diverse types of imported power plant equipment has provided the company with a wealth of experience and confidence. Along with capability to provide financing solutions, these strengths would provide the required leverage for meeting growing needs of this sector, not only in India but also abroad.

# 8

# *Voltage Control through Reactive Power Management—Case Studies*

MAHENDRA KUMAR

The role of reactive power management in voltage control is not well recognized in India and generating station/sub-station operators are usually keeping a record of voltage and reactive power flow instead of making efforts to control the voltage. Lack of financial resources for addition of capacitors, new lines or reactors is purported to be the cause of voltage problem. In real time operation, voltage level going down to 300-310 kV could be seen at some 400 kV sub-stations. At the consumers' premises, voltage goes down up to 150 V with total disregard to relevant Acts/Rules, making many of the appliances dysfunctional. High voltage is controlled by the opening of 400/220 kV lines, causing structural deficiency and associated risks in the system and increase in line losses.

We identify case studies of low voltage problem in the Southern Region (SR) and high voltage problem in the Eastern Region (ER) wherein it was noted that the proper use of reactive power capabilities of the existing generating units/sub-station equipment for reactive power management could not only improve the voltage profile but also help in substantially reducing system losses. It illustrates the methods that could be evolved for step-by-step control of low voltage or high voltage, as the case may be, along with the results thus experienced.

## INTRODUCTION

The Indian Electricity Rules prescribe limits for supply voltage at the consumers' premises at $\pm$ 6 per cent for the low or medium voltage supply and +10 per cent to 12.5 per cent for supply at extra high voltage. To meet these obligations, central planners have made a planning criteria for design of Extra High Voltage (EHV) systems to operate within ± 5 per cent for 400 kV and ± 10 per cent for 220 kV systems.

Lack of financial resources for addition of the system for better voltage control is purported to be the cause of voltage problem while lack of training of the operating personnel and seriousness of the top management to take corrective action are more responsible. Impact of low voltage on high transmission and distribution (T&D) losses also does not find adequate appreciation.

Case study of the problem of low/high voltage in the system encompassing the techno-economic aspects of the options available for voltage control in real time operation is detailed in the subsequent paragraphs. It is based on studies of low voltage problem experienced in the Southern Region and high voltage problem faced in the Eastern Region.

## THE PREVALENT PRACTICE

The majority of State Load Dispatch Centres are not monitoring the voltage, Mega Volt Ampere Reactive (MVAR) generation, MVAR flows through the lines/ICTs. Indeed monitoring of voltage is limited to a few selected points only. Truly speaking, monitoring will not be the correct word and real work is limited to the recording of the voltage at the sub-stations. Exploring the options available for voltage control through better reactive power management by making use of the reactive power capability of the existing resources is a neglected area. High frequency of 51 to 52 Hz can be seen in some of the regions with disregard to the economy of operation at the desired frequency. Low frequency going down to as low as 48 Hz is also common with total disregard to or ignorance of the regulatory mechanism. High voltage problem is more common in the systems operating at high frequency and low voltage prevailing in the systems operating at low frequency.

High rate of growth in demand and long gestation period of power systems cause gap in demand and supply due to slippage in capacity addition, funds constraints, technical, social/environmental problems and government policies. Therefore, criteria adopted at the time of planning is not applicable to real-time operation. System operation planning requires updating on a real-time basis to match the load-generation balance and other parameters prevailing at a given time.

Based on the technical analysis of low/high voltage in the system, together with the prevailing operating practices, it was noted that the existing facilities for MVAR/voltage management are not fully exploited. Adequate margin is available to control low voltage in Southern Region as well as the high voltage in Eastern Region with the existing facilities and with indirect advantages of reduced losses and better system security.

## GENESIS OF THE PROBLEM IN VOLTAGE CONTROL

Shortage of power prevailing for decades in all the regions in India made it a supplier-controlled market and the consumers had no choice but to accept the supply without questioning the quality of power. SEBs have also been working under pressure to provide the connections without caring for the economic viability of new connections or the number of hours that the existing consumers can be served. This pressure and environment left SEBs with no strength to concentrate on the technical issues for system operation. Important reasons for lack of voltage control are assessed to be as under:

(*i*) *Lack of Technical Analysis and Exploring the Available Options:* There is a lack of appreciation among system planning and operation engineers about the value of voltage control in reducing active and reactive power losses in the system, stress on equipment due to low voltage, affecting its life and damaging of the transmission system operating at lower voltages.

There is little in appreciation about the role of reactive power management in the voltage control of EHV systems. The thinking is to install capacitors to control low voltage and reactors or open some circuits on EHV lines to control high voltage. MVAR management capabilities of generators and transformer taps are grossly underutilized.

(*ii*) *Lack of Awareness among Consumers:* Indian Electricity Rules 1956 prescribes certain limits for voltage at the consumers' premises. Damage to consumers' equipment due to low as well as high voltage is quite common. However, consumers are unaware about their rights that could have helped in awakening the power utilities to make efforts towards voltage control by whatever means. Shortage of funds would have compelled them to explore the technical capability of existing resources.

(*iii*) *Regulatory Mechanism:* Voltage excursion beyond the statutory limits is an everyday affair in almost all the regions. Regulatory mechanisms are ineffective in exerting pressure on the defaulting agencies to take corrective measures.

(*iv*) *Commercial Mechanism*: There was no price tag on inter-utility MVAR exchange, resulting in lack of attention and lack of investment on MVAR management. On the consumer side, there is no price tag on domestic, irrigation and other small consumers. In case of HT/EHT consumers in the industrial category, the price tag is only one sided on to control MVAR demand that too at a very low power factor of 0.85 to 0.90 that leaves margin for high reactive power drawal.

Reactive power pricing has recently been introduced for exchange of bulk power amongst power utilities with an understanding that there will be no price tap on exchange of power when voltage at the exchange point is at 97 per cent or 103 per cent of the rated voltage.

(*v*) *Lack of Training Facilities*: MVAR management and voltage control is not covered well in the training and retraining facilities available with the utilities or the training institutes. Lack of interaction between educational institutes and the power industry is also an equally important reason.

## AREAS OF CASE STUDY, OBJECTIVE AND METHODOLOGY

Low voltage problem starts from high reactive power drawal and lack of compensation on the LT systems at the consumer end, which is further aggravated by high reactive power consumption in the transmission and distribution systems. On the other hand, high voltage problem usually starts due to reactive power generation by EHV system, mostly the 400 kV system.

The networks for the purpose of system studies developed by some agencies are limited to EHV systems (220 kV and higher voltage systems including 132 kV lines). Such networks have not been developed for the distribution systems. Given these circumstances, case studies under reference were limited to the EHV networks in the Southern and Eastern Regions.

In the multi-agency system operating in India, it is not practicable to proceed with theoretical problem definitions for true optimization. The multiple objective functions to be specified for a regional network include the following:

- To control the voltage profile;
- To minimize the need for de-energizing the transmission lines for controlling over-voltages;
- To avoid further investment in reactive power compensation devices in light of the resource crunch and expected short life of the problem;
- To keep the system losses at the minimum level;
- To maximize reactive reserves for taking care of contingencies;
- To minimize control action;
- To minimize burden on generators but make use of its full capability, when required;
- To tune the system with gap in forecast and actual load development.

While the well-known techniques of optimization by mathematical programming were traversed through, considering the developing nature of the power system, rigorous and fine analysis was not found worthwhile. There were coarse steps available based on sensitivity analysis, so a trial and error approach was adopted rather. The major problem confronted was the level of awareness within the power industry for reactive power management resulting in lack of appreciation of the study and efforts to improve the voltage and lack of cooperation from the State Power Utilities, and there is a need for improving the understanding of the role of reactive power management in system operation.

A close interaction and demonstration of the approaches to reactive power management and voltage control with visible benefits was found to be more practical. The practical constraints (technical, social and commercial) faced during this period of rapid development with resource crunch are as follows:

(*i*) Lack of concern about quality of power supply and the value of voltage control;

(*ii*) Lack of awareness about the role of reactive power management in voltage control, better capacity utilization of transmission system, reduction in transmission and distribution losses;

(*iii*) Inadequate availability of data, poor telemetering, particularly for reactive power flow and voltage;

(*iv*) Reluctance on the part of the sub-station operators to change the taps of transformers even though they were fitted with on load tap changers. This was partly due to technical ignorance and partly due to the fear of technical problems in tap changing;

(*v*) Hesitation in utilizing the reactive power generation/absorption capability of the generators even within their design capability;

(*vi*) Undermining the role of integrated EHV network in reactive power generation/absorption and keeping a number of 220 kV lines open after commissioning 400 kV lines;

(*vii*) Scheduling and dispatch activities limited to active power;

(*viii*) Delegation of the responsibility for reactive power control and regulation amongst the pool members.

## CASE STUDY I—LOW VOLTAGE PROBLEM IN THE SOUTHERN REGION

At the time of the case study, with peak demand of about 12,000 MW, agriculture pumps load constituted about 30 per cent of the total demand in Southern Region. Reactive power requirement of small size, and inefficient induction motors in such pumps was quite high. In integrated system operation, voltage control was taken as no man's responsibility and voltage level of 295 to 300 kV on the 400 kV bus was quite common at some important points like the capital cities of Chennai and Bangalore.

Telemetered data for reactive power flows was not in practice. This data was recorded in local log books only at select power stations and sub-stations. Grid map of the Southern Region showing the 400 kV central transmission system and 220 kV inter-state lines is given in Fig. 8.1. The following steps were practised to control voltage:

- Switching off the bus reactors;

- Installation of capacitors at distribution points (lagging far behind the schedule of activities);
- Changing the transformer taps on ICTs to extremes to improve the voltage on the lower side.

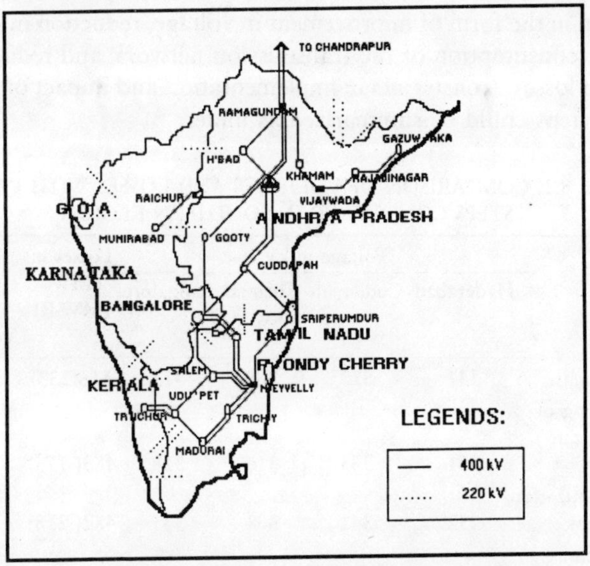


FIGURE 8.1: GRID MAP OF SOUTHERN REGION

A study was undertaken in February 1994 to explore possibilities of improvement in voltage on an immediate basis without additional load restrictions and with the existing resources. This was extended to find the line of action for further improvement with minimum investment.

With the above objectives, the following four steps were considered:

(*i*) Removal of line reactors at selected points;
(*ii*) Change of transformer taps on 400/220 kV ICTs;
(*iii*) Closing of 220 kV inter-state lines that were open;
(*iv*) Reduction in reactive power drawal by SEBs at selected points by increasing the reactive power generation at the load centre based power stations or installation of capacitors at some selected points.

Results of a study to assess the impact of the above steps is given in Table 8.1. Studies were carried out with system parameters nearest to those prevailing at that time. These included all 400 kV and 220 kV transmission systems and a major part of the 132 kV system, except some 132 kV lines in radial mode. Multiple benefits could be seen in the form of improvement in voltage, reduction in reactive power consumption of the transmission network and reduction in system losses. Constraints in implementation and impact of various other steps could be summarized as under:

TABLE 8.1: COMPARISON OF VOLTAGES AND LOSSES WITH VARIOUS STEPS CONSIDERED IN SOUTHERN REGION

| Case | Voltage in kV | | | | Losses in MW (MVAR) | Reduction in losses MW (MVAR) |
|---|---|---|---|---|---|---|
| | Hyderabad | Cuddapah | Chennai | Bangalore | | |
| 1. Base case | 347 | 323 | 309 | 313 | 516(233) | – |
| 2. Closing of interstate lines | 351 | 335 | 316 | 328 | 483(-176) | 33(409) |
| 3. Optimization of taps | 352 | 341 | 331 | 335 | 482(-218) | 34(451) |
| 4. Reactor removal at Trichur and Udumalpet | 348 | 326 | 311 | 316 | 509(127) | 7(106) |
| 5. Reduction of MVAR at Bangalore, Hyderabad, Chennai and Cuddapah | 371 | 358 | 344 | 349 | 418(-933) | 98(1,166) |
| 6. Steps 2 to 4 | 355 | 350 | 339 | 345 | 465(-520) | 51(753) |
| 7. Steps 2 to 5 | 376 | 377 | 370 | 370 | 399(-1371) | 117(1604) |

### Closing of Interstate Lines

After the commissioning of the 400 kV Central Transmission System in the late 1980s, the SEBs opened almost all the 220/132 kV lines assuming that adequate 400 kV system is available for inter-connection of various systems within the region. The central planners had, in fact, approved the construction of a 400 kV system on the presumption that all the existing 220/132 kV lines would continue to be in service.

At the time of study in February 1994 out of a total of 12, four interstate 220 kV lines could still be put into service. The other lines necessitated repairs before charging.

System studies revealed that substantial improvement could be made in the voltage profile and reduced losses by 33 MW and 409 MVAR by this step. However, SEBs did not agree to close the lines and asked for more time.

### Optimization of Transformer Taps

Due to the reluctance of the majority of the SEBs to change the taps from extreme positions, this exercise had to be limited to select 400/220 kV ICTs. In the system study package, the impact of changes in the tap position of ICTs at various locations was seen step by step and appropriate positions were selected. There was significant improvement in voltage (of the order of 20 kV at critical points) and reduction in losses to the order of 34 MW and 451 MVAR. The impact could have been much larger if there optimization of taps at all places.

### Review of On-line Reactive Compensation

In the Southern Region, on-line reactors are found only in 400 kV systems. An average of 43.60 per cent compensation was in position. Removal of reactors from long lines was not possible on account of dynamic over-voltage problem. Since most of the lines remained heavily loaded round the clock and redundancy was not available, it was also not possible to remove the reactors for select hours on a daily basis. There was a wide difference of opinion on the minimum required on-line compensation and selection of reactors to be removed. 3 × 50 MVAR reactors from short-length lines near the load centres were removed leaving 39.30 per cent on-line compensation, with the provision that they could be put back into service during light load seasons, if found necessary. There was a marginal improvement in voltage and reduction in losses to the line of 7 MW and 106 MVAR.

## Reduction of MVAR Demands at Select Points

Under the conditions prevailing in February 1994, reduction in MVAR drawal at critical points was possible by measures like increase in MVAR generation at power stations located near the load centres, use of synchronous condensers already installed, staggering of MVAR sensitive load, proper maintenance of the existing capacitors and installation of new capacitor banks. For estimation purposes, studies were conducted to see the impact of reduction in MVAR drawal to 50 per cent of that prevailing at that time. The results were astonishing. Voltage improved by about 30 kV and losses reduced by 98 MW and 1166 MVAR.

This step could be implemented partially as two SEBs agreed to increase reactive power generation at some select generating stations.

With the steps proposed above, involving no investment and no additional restrictions, it was possible to improve the voltage by 50 to 60 kV and reduce losses by more than 100 MW. However, these steps could be implemented only partially due to resistance from some of the constituents.

Voltage at most critical points could still be improved by 30 to 40 kV during the period of experiment for about 10 days. Minimum voltage levels at important 400 kV sub-stations at the beginning, at the mid-point and at the end of the experiment are given in Table 8.2.

TABLE 8.2: MINIMUM VOLTAGES AT VARIOUS 400 KV SUB-STATIONS IN SOUTHERN REGION DURING THE PERIOD OF EXPERIMENTATION

| | Date | | | | | |
|---|---|---|---|---|---|---|
| S/S | 01.02.94 | 02.02.94 | 06.02.94 | 07.02.94 | 12.02.94 | 13.02.94 |
| Hyderabad | 340 | 350 | 352 | 355 | 360 | 370 |
| Nagarjuna Sagar | 360 | 357 | 365 | 365 | 375 | 375 |
| Gooty | 325 | 325 | 340 | 342 | 350 | 352 |
| Cuddapah | 310 | 310 | 330 | 330 | 355 | 355 |
| Bangalore | 300 | 295 | 316 | 319 | 340 | 350 |
| Chennai | 295 | 300 | 330 | 320 | 340 | 340 |
| Salem | 315 | 315 | 335 | 338 | 360 | 360 |
| Udumalpet | 312 | 330 | 340 | 340 | 370 | 370 |
| Trichur | 315 | 310 | 340 | 340 | 360 | 360 |
| Madurai | 320 | 320 | 345 | 340 | 370 | 370 |

The rate of load growth in the Southern Region is much higher than the increase in the generating capacity and other infrastructure. This has put extra pressure on the grid, pulling back the voltage to lower levels. There has been an improvement in the voltage level at some critical points after the commissioning of some new 400 kV lines in the nearby area. However, the addition of new lines is cost consuming and not the correct answer as it will improve the voltage at select points and not in the entire grid.

## CASE STUDY II—HIGH VOLTAGE PROBLEM IN THE EASTERN REGION

The timely completion of the super grid transmission system and slippages in the commissioning of generating stations, together with retarded growth in demand and the addition of captive generation by major consumers at load centres left the system grossly under loaded and consequentially causing a high-voltage problem. Central Sector 400/220 kV and other 400 kV transmission systems in the Eastern Region existing at the time of case study in February 1995 are shown in Fig. 8.2.

Over-voltage protection in the 400 kV system is set at 1.1. To contain the voltage within this limit, it was common practice to open some of the 400/220 kV transmission lines, causing structural deficiency and high risk to system security. The frequency of opening these lines was to the order of 80 per month merely in the central transmission system.

During the case study, multiple objectives were selected as under:

- To keep the voltage profile within acceptable limits;
- Minimize further investment;
- Minimize the need for opening the lines affecting the system security;
- Minimize burden on generators but make use of MVAR capability, when required;
- Keep the system losses at the minimum level.

To start with, the steps taken were more by trial and error as the SEBs were interested in seeing the results and the desired level of participation was not forthcoming. There had been demands for

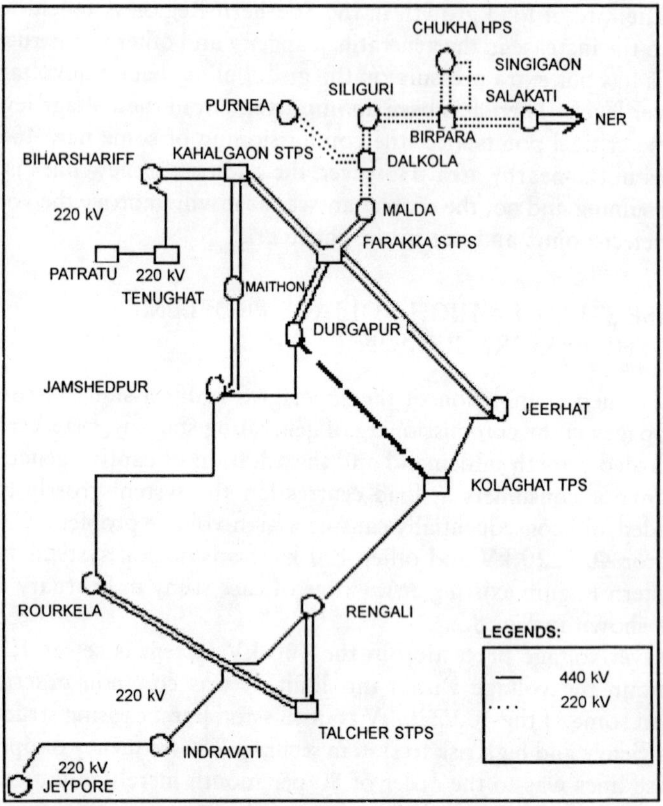


FIGURE 8.2: CENTRAL SECTOR AND 400 kV SYSTEM IN EASTERN REGION

installation of more reactors. In the developing phase, it was felt that the load would grow within two to three years and it may not be worth opting for huge investment. The following options were explored:

### Switching Off Reactors/Capacitors

Awareness was created among the operators to ensure switching off the capacitors and switching on the reactors near to the points facing high voltage before opening any line. Since most of the bus reactors were on the central lines, the response was positive.

## Tap Staggering

Since over voltage is a continuing problem and likely to continue for some time, the creation of artificial reactors through tap staggering was considered at such sub-stations where transformers always remained under-loaded. This step had been taken at Durgapur (West Bengal), Rengali (Orissa) and Maithan (Jharkhand). The results of such staggered operation at Durgapur are given in Table 8.3. An artificial reactor of 25 MVAR could be created without investment.

TABLE 8.3: TAP STAGGERING AT DURGAPUR

| Auto Transformer | Before Staggering | | | After Staggering | | |
|---|---|---|---|---|---|---|
| | | Losses | | | Losses | |
| | Tap Position | MW | MVAR | Tap Position | MW | MVAR |
| 1 | 16 | 0.3 | 6.01 | 16 | 0.78 | 15.64 |
| 2 | 16 | 0.29 | 5.87 | 16 | 0.77 | 15.49 |
| 3 | 16 | 0.29 | 5.8 | 6 | 0.57 | 11.46 |
| Total | | 0.88 | 17.68 | | 2.12 | 42.59 |

Tap staggering was also done in loops formed by 400 kV and 220 kV lines between Farakka and Malda in West Bengal. A reactor of 22 MVAR could be created at that location.

## Reactive Power Absorption by Generators

Many of the power stations were generating reactive power even under conditions of high voltage in the area. Old thermal stations could be persuaded to reduce reactive power generation but did not agree to absorb the same when required. Many of the stations with 200 and 500 MW units could be persuaded to go to the leading side but not up to the extreme values as per capability curve.

Hydro generators could be used better for reactive power absorption. Chukha Hydro Project with 4 × 84 MW machines could be persuaded to absorb up to 30 to 35 MVAR on each machine. Even if three machines are operating, an absorption of about 100 MVAR helped in control of over-voltage in that area. These machines were also tested to operate as synchronous motors absorbing about 40 MVAR each.

### Dropping of Lines

Despite partial support from the grid users, there are occasions when voltage goes high even after taking the above steps. This is more so during the rains when the agricultural load with high reactive power component goes down and some thermal units are taken out for annual maintenance.

### Captive Generators

The eastern grid is operating with a number of captive generators owned by core sector industries like steel and aluminium. They could also contribute in reactive power management through their generators and capacitors. Commercial and contractual aspects are coming in the way as far as optimum utilization of such capacity is concerned. Avenues have been opened for the utilities to take necessary steps for their participation in grid management.

The steps discussed above helped in the control of over-voltage by a combination of options rather than exclusively by the opening of lines. Fig. 8.3 shows the number of openings of 220/400 kV lines in the super grid for a few months before the above steps were taken (February 1995) and a few months after that. Incidence of

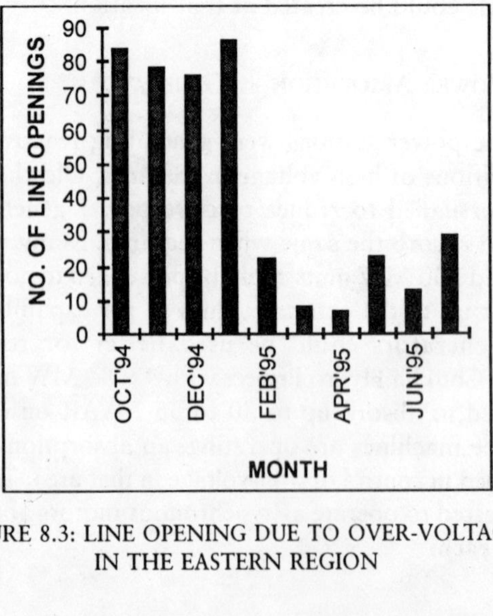


FIGURE 8.3: LINE OPENING DUE TO OVER-VOLTAGE IN THE EASTERN REGION

opening of lines came down from about 80 per month to about 20 per month.

## RECENT INITIATIVES IN COMMERCIAL MECHANISM/REGULATORY CONTROL

The Indian Electricity Grid Code (IEGC) issued by Central Electricity Regulatory Commission (CERC) makes provision for commercial mechanism as well as regulatory control on exchange of reactive power amongst bulk power utilities. It specifies that in case the voltage at the exchange point is below 97 per cent, the utility trading captive power will be required to pay for such charge. In case the voltage is above 103 per cent, the bulk power utility gets paid for drawal of captive power. The tariff for the exchange of such power has been fixed at 4 paise per kWh up to 31 March 2001, to be escalated at the rate of 5 per cent per year therefrom, unless revised by CERC. The system has been introduced in various regions excluding the north-east, between July 2002 and April 2003.

The Act further provides that switching in/out of all 400 kV bus and line reactors and tap changing on all 400/220 kV ICTs shall be done as per the instructions of Regional Load Dispatch Centres (RLDC). In respect of generating stations, the Act specifies that Central Generating stations-generated power shall be absorbed as per the instructions of RLDCs within the capability limit of respective generating units and no payment shall be made to the generating company for such VAR generation/absorption.

It is expected that the recent initiatives in the introduction of commercial mechanism in regulatory control on exchange of reactive power would help improve reactive power management and give technical and financial benefit to power utilities.

## CONCLUSION

Regional grids in India are operating at voltage levels, much beyond the statutory limits prescribed by the relevant Acts and Rules. Technical analysis and the results of steps taken in the Southern Region and the Eastern Region give enough evidence of the scope for improvement in solving low-voltage as well as high-voltage problems. It is more important to record that this can be done by better use of the existing resources and without further investment.

However, there is an urgent need for the top managers to realize their responsibility of honouring the regulations as well as ensuring quality power supply to the consumers. Extensive training is necessary at the operators' level to make them sensitive to grid parameters and making use of the capabilities built in the existing generation and transmission equipment.

System behaviour in the Northern Region is a combination of low and high voltage problem. This region incorporates a long haulage of quantum power, making the behaviour of reactive power highly load sensitive. Load and its power factor in the Northern Region is also highly weather sensitive. A combination of experiences in the Southern and Eastern Regions could be implemented in to address the high-voltage as well as the low-voltage problems.

Since a major part of the generation and transmission system is owned and operated by the vertically integrated SEBs, the extent of their participation in grid management will be a measure of success. Apart from exploring the options available, more stress will have to be given on increasing the awareness and training of the engineers to appreciate the role of reactive power management in voltage control and other indirect benefits. The price tag on reactive power exchange and reactive power support by generating companies may further expedite the process of voltage control. Indeed efforts in this direction will have to be a continuous process.

Above all, reactive power management can help in substantial reduction of transmission and distribution losses. Whereas the case study in the Southern Region was limited to the EHT system, reduction in losses will be far greater in the HT and LT systems as any improvement in the voltage in the EHT system will automatically improve the voltage in HT and LT systems. Thus, in addition to improving the quality of power supply and reducing the investment in capacitors/reactors, SEBs will be getting substantial financial gain in terms of reduced losses in the transmission and distribution networks.

# 9

## *Transmission Line Congestion Management Using Bid-Areas Division Technique*

D.P. KOTHARI AND PARUL GOYAL

It is well known that transmission line congestion is one of the drawbacks of the deregulated power sector. Developing countries like India have non-uniform percentage power deficiency (PPD) in all the regions. PPD is defined as the percentage difference of power met and power demand in a certain region. Transmission line congestion occurs because the load requirement and hence the power transactions over the lines increase in some regions due to higher PPD. In this paper, a 'bid-areas division' technique of transmission line congestion management is developed under modern deregulated power environment. In this technique, to protect the network from congestion PPD difference is minimized for all the regions throughout the country, which distributes transmission loading uniformly.

### INTRODUCTION

'Deregulation' implies reformation of the rules and regulations formed by the government for the traditional power sectors.[1] According to these rules, only one power company is possible in a certain area, which has a fixed price of electricity. In the deregulated environment, many Independent Power Producers (IPP) are possible in a certain area and these IPPs can set their own electricity prices for maximizing their benefits. An electricity market is formed in

*The authors are thankful to the Central Electricity Authority for providing the latest power data on India.

which IPPs and power consumers submit their price-bids to an independent personality known as the Independent System Operator (ISO), who matches the submitted price-bids by applying a suitable bid-matching technique to calculate a market-clearing price.

In a deregulated power structure, transmission network remains regulated or only one transmission network is desirable in a certain area because it occupies a huge space and involves high capital cost. All IPPs use this transmission network for their power transactions. Consequently, a situation of overloading or congestion may occur in the transmission network, resulting in increased transmission losses and instability of the network. Hence, there is a need to develop a congestion management technique to remove the transmission line congestion in the initial stages.

In developing countries like India percentage power deficiencies in all the regions are not same, as shown in Table 9.1. The difference of peak demand and the demand met in each region of India is negative. Northern, southern and eastern regions have lesser power deficit than the western and north-eastern regions.

Table 9.2 shows the monthly monitored generating capacity (MW) in India for the month of February 2004. Regional tariffs are also different in India, but not fixed under a deregulated environment. This non-uniformity of power generation, electricity prices and its consumption in all the regions of a developing country encourages excess power transactions across regional boundaries of the country. Hence, an interconnected power network is developed, which sometimes overloads the transmission lines due to high power demand at cheaper prices from a particular generating place.[2]

To avoid the transmission network from congestion, a 'congestion management' technique is suggested in this paper, which gives better results than the other congestion management techniques.

TABLE 9.1: DIFFERENCE OF PEAK DEMAND AND PEAK MET IN INDIA FOR FEBRUARY 2004[3]

| Region | Peak Demand | Peak Met | Surplus/Deficit | |
|---|---|---|---|---|
| | (MW) | (MW) | MW | % |
| All India | 82,575 | 73,077 | –9,498 | –11.50 |
| Northen | 22,724 | 21,538 | –1,186 | –5.20 |
| Western | 28,334 | 22,809 | –5,525 | –19.50 |
| Southern | 22,678 | 20,724 | –1,954 | –8.60 |
| Eastern | 7,708 | 7,077 | –631 | –8.20 |
| North-Eastern | 1,131 | 929 | –202 | –17.90 |

TABLE 9.2: MONTHLY MONITORED CAPACITY IN INDIA[4]

| Type of Generation Monitored Capacity | MW |
|---|---|
| I. All India | |
| (*i*) Thermal | 76,670.10 |
| (*ii*) Nuclear | 2,720.00 |
| (*iii*) Hydro | 27,950.20 |
| (*iv*) Total | 1,07,340.30 |
| II. Regions | |
| 1. Northern | |
| (*i*) Thermal | 18,869.70 |
| (*ii*) Nuclear | 1,180.00 |
| (*iii*) Hydro | 9,584.80 |
| (*iv*) Total | 29,634.50 |
| 2. Western | |
| (*i*) Thermal | 25,361.50 |
| (*ii*) Nuclear | 760.00 |
| (*iii*) Hydro | 4,529.80 |
| (*iv*) Total | 3,0651.30 |
| 3. Southern | |
| (*i*) Thermal | 17,202.40 |
| (*ii*) Nuclear | 780.00 |
| (*iii*) Hydro | 10,382.50 |
| (*iv*) Total | 28,364.90 |
| 4. Eastern | |
| (*i*) Thermal | 14,120.00 |
| (*ii*) Nuclear | 0.00 |
| (*iii*) Hydro | 2,372.40 |
| (*iv*) Total | 16,492.40 |
| 5. North Eastern | |
| (*i*) Thermal | 1,116.50 |
| (*ii*) Nuclear | 0.00 |
| (*iii*) Hydro | 1,080.70 |
| (*iv*) Total | 2,197.20 |

## CONGESTION MANAGEMENT TECHNIQUE

In this section, the new congestion management technique is compared with the old technique.

### Old Technique of Congestion Management

In the old technique of congestion management, the ISO divides all the regions of a country into several parts on the basis of congestion intensity present in each region. Higher congestion intensity areas are separated from the lower congestion intensity areas. The ISO informs to all the transmission line users about the network congestion and the divided areas. Users submit their price-bids again, different for each area. Low price-bids for the higher congestion intensity areas are rejected. Different MCPs are set for each area by a bid-matching technique in the electricity market. MCPs for the higher congestion intensity areas are kept higher to the MCPs for the lower congestion intensity areas. The results of this technique are discussed below.

The increase in MCPs at the higher congestion intensity areas affects lower congestion intensity areas too, because the consumers of the higher congestion intensity areas prefer to set contracts with the IPPs of the lower congestion intensity areas to get power at lower MCPs. Consequently, the load in the lower congestion intensity areas increases, resulting in a high load-generation imbalance in these areas. Although, in the higher congestion intensity areas transmission lines' overload is removed due to their load shift to the lower congestion intensity areas, the generating companies at these bid-areas bear a great loss.

If the ISO lowers the price of electricity in these areas in the normal situation to restore the load, then again network congestion may occur. Hence, congestion management will be a never-ending process.

## New Technique of Congestion Management

In developing countries like India, PPDs are not the same for all the regions, e.g. northern, southern and eastern regions of the country have lower PPD than the western and north-eastern regions. Hence, the power requirement at higher PPD areas is greater to the lower PPD areas. The consumers of higher PPD areas set contracts with more number of power suppliers than the lower PPD areas. The supply of a large amount of power to one region creates transmission line overloading at that place. The new technique of congestion management deals with this situation. In this technique, all the regions of the country are divided into several parts known as 'bid-areas' on the basis of their PPD, if the network congestion is expected or identified.

To identify the overloading of transmission lines, the ISO monitors the line flow and maximum power transfer limit (MPTL) of each line in the transmission network by conducting a load flow analysis either manually or by using a computer tool. Here, a computer software named *Power Simulator* is used to simulate the complex 14-bus transmission network to obtain line flow and MPTL of each line. If the line flow exceeds its maximum transfer limit, then ISO declares the expected network congestion.

Once the ISO identifies the expected network congestion, he/she divides all the regions into several areas known as 'bid-areas' on the basis of their PPD. Higher PPD areas are separated from the lower PPD areas to form bid-areas.

The ISO informs to all the transmission network users about the network congestion and the divided bid-areas. All the users submit new price-bids, different for each bid-area, to the ISO in the market place. The ISO rejects low price-bids for the higher PPD bid-areas. Different MCPs are set by a bid-matching technique for all the bid-areas in the electricity market. The MCPs are kept lower for the lower PPD bid-areas and higher for the higher PPD bid-areas.

Increment in the MCPs at the higher PPD bid-areas discourages consumers to set contracts with the IPPs of these bid-areas. The results of this technique are given below:

- The load from the higher PPD areas decreases and their PPD improves.
- The differences of PPDs for all the bid-areas are minimized throughout the country, which maintains almost uniform loading of the transmission network for all the regions.
- High MCPs at the higher PPD areas encourages the IPPs to set up their generating plants in these bid-areas to earn more.

The advantages of this technique are given below:

- Higher MCPs at the higher PPD encourages IPPs to install their new generating plants at these bid-areas.
- If a congestion fee similar to the old technique is charged to all the users, then it penalizes large as well as small power users. In the new technique, the whole market price structure is changed, which affects only the large power users, responsible for the transmission network overloading.
- Users having their power transactions over a small distance across a zonal boundary are given price concessions.

## TRANSMISSION POWER MONITORING TOOL

Transmission line network is very complex in structure. Hence, it is very difficult to conduct load flow analysis manually for this complex network. The Software *Power Simulator* is used to compute line flow and the MPTL of each line by load flow analysis. It is an easier and faster way to determine line flow in the transmission network, which facilities the work of the ISO. Without fast monitoring of the transmission line flow and maximum transfer limit, this suggested technique of congestion management is difficult to implement.

### CASE STUDY

Consider a 14-bus Institute for Electrical and Electronics Engineers (IEEE) model of transmission network with 20 branches as shown in the Fig. 9.1. Branch data are given in the Table 9.3. The following results are obtained by conducting load flow analysis with *Power Simulator* software:

(*i*) Y-matrix
(*ii*) Bus data
(*iii*) Power flow data

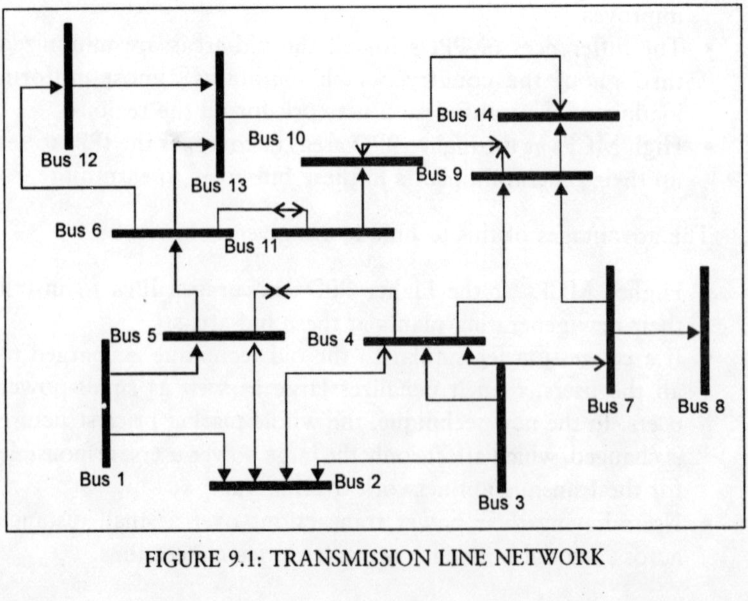


FIGURE 9.1: TRANSMISSION LINE NETWORK

TABLE 9.3: BRANCH DATA

| From | To | Resistance (pu) | Reactance (pu) | Line Limit (MVA) |
|---|---|---|---|---|
| 1 | 2 | 0.019380 | 0.059170 | 300 |
| 1 | 5 | 0.054030 | 0.223040 | 100 |
| 2 | 3 | 0.046990 | 0.197970 | 100 |
| 2 | 4 | 0.058110 | 0.176320 | 100 |
| 2 | 5 | 0.056950 | 0.173880 | 100 |
| 3 | 4 | 0.067010 | 0.171030 | 100 |
| 4 | 5 | 0.013350 | 0.042110 | 80 |
| 4 | 7 | 0.010000 | 0.209120 | 100 |
| 4 | 9 | 0.015000 | 0.556180 | 100 |
| 5 | 6 | 0.01085 | 0.252020 | 100 |
| 6 | 12 | 0.122910 | 0.255810 | 100 |
| 6 | 13 | 0.066150 | 0.130270 | 100 |
| 6 | 11 | 0.094980 | 0.198900 | 100 |
| 7 | 8 | 0.013546 | 0.176150 | 100 |
| 7 | 9 | 0.01657 | 0.110010 | 100 |
| 9 | 10 | 0.031810 | 0.084500 | 100 |
| 9 | 14 | 0.127110 | 0.270380 | 100 |
| 10 | 11 | 0.082050 | 0.192070 | 100 |
| 12 | 13 | 0.220920 | 0.199880 | 100 |
| 13 | 14 | 0.170930 | 0.348020 | 100 |

In the suggested technique, line flow is compared with the power carrying capacity of each transmission line. Hence, we are listing power data only as obtained by conducting load flow analysis with Power Simulator (see Table 9.4).

TABLE 9.4: POWER FLOW DATA

| | | |
|---|---|---|
| 1 to Bus 2 | 157.06784 MW | –17.48119 MVAR |
| 1 to Bus 5 | 75.90918 MW | 4.29688 MVAR |
| 2 to Bus 5 | 41.46810 MW | 0.16220 MVAR |
| 2 to Bus 4 | 56.38597 MW | –1.36277 MVAR |
| 2 to Bus 3 | 73.20470 MW | 5.95497 MVAR |
| 2 to Bus 1 | –152.75996 MW | 30.63377 MVAR |
| 3 to Bus 4 | –23.31706 MW | 3.46381 MVAR |
| 3 to Bus 2 | –70.88348 MW | 3.82436 MVAR |
| 4 to Bus 5 | –62.96768 MW | 11.18207 MVAR |
| 4 to Bus 9 | 16.54057 MW | –0.66102 MVAR |
| 4 to Bus 7 | 29.63982 MW | –10.57618 MVAR |
| 4 to Bus 2 | –54.69314 MW | 6.49924 MVAR |
| 4 to Bus 3 | 23.68209 MW | –2.53215 MVAR |
| 5 to Bus 6 | 42.60739 MW | –1.83420 MVAR |
| 5 to Bus 1 | –73.12946 MW | 7.17802 MVAR |
| 5 to Bus 2 | –40.57130 MW | 2.57592 MVAR |

(contd.)

TABLE 9.4 (contd.)

| | | |
|---|---|---|
| 5 to Bus 4 | 63.49202 MW | -9.52815 MVAR |
| 6 to Bus 11 | 6.38435 MW | 1.71594 MVAR |
| 6 to Bus 12 | 7.61189 MW | 2.29621 MVAR |
| 6 to Bus 13 | 17.22580 MW | 6.28972 MVAR |
| 6 to Bus 5 | –42.41936 MW | 6.20163 MVAR |
| 7 to Bus 8 | 0.06789 MW | -23.14941 MVAR |
| 7 to Bus 9 | 29.47864 MW | 10.58480 MVAR |
| 7 to Bus 4 | -29.54472 MW | 12.56504 MVAR |
| 8 to Bus 7 | –0.00092 MW | 24.02040 MVAR |
| 9 to Bus 10 | 6.18526 MW | 6.04615 MVAR |
| 9 to Bus 14 | 10.14637 MW | 4.79730 MVAR |
| 9 to Bus 4 | –16.50109 MW | 2.12460 MVAR |
| 9 to Bus 7 | –29.32865 MW | -9.58903 MVAR |
| 10 to Bus 11 | –2.83845 MW | 0.18207 MVAR |
| 10 to Bus 9 | -6.16264 MW | –5.98605 MVAR |
| 11 to Bus 10 | 2.84485 MW | –0.16709 MVAR |
| 11 to Bus 6 | –6.34520 MW | –1.63397 MVAR |
| 12 to Bus 13 | 1.44011 MW | 0.54488 MVAR |
| 12 to Bus 6 | –7.53862 MW | –2.14373 MVAR |
| 13 to Bus 14 | 4.94841 MW | 0.61304 MVAR |
| 13 to Bus 6 | –17.01604 MW | –0.54028 MVAR |
| 13 to Bus 12 | –1.43503 MW | –0.54028 MVAR |
| 14 to Bus 9 | -9.99416 MW | –4.47352 MVAR |
| 14 to Bus 13 | –4.90683 MW | –0.52838 MVAR |

Line flow data are displayed on the ISO's monitor. If the monitored line flow exceeds the maximum transfer limits of the lines, then the ISO declares expected network congestion.

## CONCLUSION

In the modern deregulated transmission network, congestion management is essential for the stable operation of the network. In some congestion management techniques, regions of a country are divided into several parts on the basis of the congestion intensity present in each region. In the market place, higher MCPs are set for higher congestion intensity areas and lower MCPs are set for lower congestion intensity areas. This technique is not successful because the load at the higher MCP areas is shifted to the lower MCP areas, resulting in a high generation-load imbalance in these areas.

The technique of congestion management developed in this paper is based upon the bid-areas division on the basis of the PPD in each region of a country. Lower MCPs are set for the lower PPD bid-areas and higher MCPs are set for the higher PPD bid-areas. Consequently, some consumers from the higher MCP bid-areas are shifted to the lower MCP bid-areas. Hence, by applying this technique the differences of PPDs for all the regions are minimized, which distributes the transmission network loading almost uniformly in all the regions.

When the transmission line congestion is removed, all bid-areas can be united again to form regions and market-clearing prices are set for these regions by matching new price-bids submitted by the IPPs and the customers. Due to increased generation in some regions congestion may not occur in these regions again.

Transmission network has a complex structure and it is difficult to conduct load flow analysis manually. Hence, software installation is needed for load flow analysis for faster and better response and to facilitate the work of the ISO. However, installation of the computer software is expensive and the ISO must be an expert in handling this software.

## NOTES

1. Lorrin Philipson and H. Lee Willis, *Understanding Electric Utilities and Deregulation*, Marcel Dekker, Inc., 1999; Kankar Bhattacharya, Math H.J. Bollen, Jaap E. Daalder, *Operation of Restructured Power Systems*, The Kluwer International Series in Engineering and Computer Science, vol. 621, Kluwer, Boston, USA, 2001.
2. A. Wood and B. Wollenberg, *Power Generation, Operation, And Control*, New York: John Wiley and Sons, 1984; D.P. Kothari and I.J. Nagrath, *Modern Power System Analysis*, 3rd edn., New Delhi: Tata McGraw Hill, 2003.
3. Central Electricity Authority, Ministry of Power, Government of India, Power Sector Data 2003 and 2004.
4. Ibid.

# 10

## *Integrated Resource Planning in Supply Side Management in Power System*

PREM K. KALRA, YOGESH K. BICHPURIYA
AND V.P. SINGH

Ongoing changes in the structure and regulation of the power industry require changes in approach to resource planning. Given the potential for commodity markets (both natural gas and electric) to exhibit price swings or volatility, alternative resource plans must be evaluated in terms of their exposure to this volatility, in addition to their long-run average costs. Furthermore, unpredictability in the future costs of new supply alternatives arising from fuel cost (primarily natural gas price) provides analysis leading to a comprehensive portfolio and strategy for supply acquisitions, transmission investments and demand side management along with the consideration of environmental issues. The purpose of Integrated Resource Planning (IRP) is to provide reliable, safe and least cost electric service to customers. This paper pays emphasis on supply side management and its integrated resource planning.

### INTRODUCTION

The Indian power sector has been increasing its installed capacity; it has risen from 30,000 MW in 1981 to over 1,00,000 MW in 2001. Despite this growth in supply, its power systems are struggling to overcome chronic power shortages and poor power quality. With demand exceeding supply, severe peak (around 18 per cent) and energy (around 10 per cent) shortages continue to plague the sector. Shortages are exacerbated by inefficiencies in power generation, distribution and end-use systems. The inefficiency on the supply side is due to unavailability of better quality of natural resources, less use of newer technology and lack of awareness. The inefficiencies in the end-use systems are due to irrational tariffs, technological obsolescence of industrial process and equipment, lack of awareness, and inadequate policy drivers (such as energy efficiency standards and labelling system and financial incentives) in India.

As the deregulation of the electric power industry progresses in India, the basic structure of power systems will change drastically. For example, it is expected that small-scale distributed generation facilities (such as photovoltaic, fuel cell generation) and dispersed type energy storage systems (such as secondary battery, electric vehicle and SMES) are installed in the demand side or the power distribution systems. The introduction of a disaggregated, competitive electricity supply industry will affect the appropriateness of various planning models currently used. One such approach that has been significant in promoting energy efficiency and demand side management in a number of countries is IRP.

IRP is a planning process that seeks the least cost option for meeting customers' electricity demands. In determining the least cost option, IRP evaluates all supply and demand side options over a forecast period from a societal perspective. It implies significant regulatory oversight which can be applied by a number of means. In its broadest context IRP includes all energy forms and the whole national energy market. It can also be implemented for the electricity industry, or at an individual utility level. A key feature is the regulatory oversight of the process. It can take a variety of forms; at one extreme, government may exert significant control over the process. This may involve establishment of an independent energy planning body with responsibilities including:

- Development of IRP procedures, including costing of environmental externalities;
- Oversight of public consultation and review of plans;
- Amendment of plans;
- Setting and reviewing utility performance criteria identified in the plans;
- Reviewing and approving major capital investment decisions.

At the other extreme, a utility may simply be required by government to periodically publish report on its strategic planning process, and seek public comments.

IRP may be initiated in two main ways:

- By governments applying IRP across one or more major industry sectors—this could be termed as 'mandatory IRP';
- By individual electricity businesses, applying IRP only within that particular business—this could be termed as 'business-related IRP'.

Whatever the form of the IRP, the following elements are common:

- It is a continuing and iterative process, ideally planning, implementation and evaluation;
- It is an attempt to balance long- and short-term goals, and to meet multiple goals;
- All resource options are compared in the one forum and a portfolio approach is promoted;
- Public discussion and debate are involved.

Mandatory IRP works well in an electricity industry composed of vertically integrated monopoly electricity businesses. In an industry structure where the functions of the electricity industry have been unbundled into separate businesses, both mandatory and business-related IRP can still be undertaken in the natural monopoly elements of the industry, i.e. the transmission and distribution network sectors.

Many countries are looking for 'deregulating' or 'restructuring' their electric power industries. The basic purpose of this restructuring is to open electric power industries to competition among suppliers. These efforts are considered to provide better quality of electric supply to consumers at a lower price. The point to consider is how the IRP process will include social benefits with the deregulation of the electric power industry.

The electricity planning problem in many developing countries is more complicated as compared to developed countries. Generation of power is difficult because electric power industries require large capital investment. The developing countries have a problem to substitute investment in cost-effective reductions in electricity use for investments in power plants. The combined planning for increasing electricity capacity to meet growing demand with cost-effective reductions in use of electricity is called IRP. The growing demand for electricity throughout the world is met by electric utilities by building more power plants. Increased burning of fossil fuels in these power plants could pollute the atmosphere. Burning of fossil fuels would produce more $CO_2$ which contributes to global warming. In developing countries, due to high growth in power generation, the $CO_2$ emissions grew fastest in last few years. Over the last two decades $CO_2$ emission has increased by about 80 per cent in these countries. The combination of resources different from those used in the past and the integration of these resources in a cost effective way could ease the health, economic, and environmental effects of increased demand for electric power.

Environmental problems such as emissions of $CO_2$ gas from the power plants are more complex while building power plants to meet the increasing demand for power. According to the environment regulatory bodies, countries should encourage programmes to limit the greenhouse gas emissions.

IRP process may alter the pattern and the levels of electricity demand and construction of non-conventional power plants. IRP takes the future assessment of energy supply and demand. It combines them while considering environmental and social factors, energy efficiency and load management programmes, and different limitations of the resources. Technically it should treat the demand side resources and the supply side resources equally.

Practically, IRP process works as under:

- It analyses the present situation of existing resources, future demand for electricity and identifies the reliability of service, environmental impacts and economical price for supply.
- It identifies resource options that would compensate for discrepancies between expected loads and existing capacities.
- The options are evaluated considering the economic, social and environmental factors.
- The resources are combined to meet the demand for electricity in a way that satisfies all regulatory issues and avoids risks.
- The plan, thus obtained, is implemented.

## HOW TO DO IRP IN INDIA

On the supply side, the elementary problem faced by the power sector is the poor financial condition of State Electricity Boards

(SEB) or successor entities in most states. Over the years SEBs have been causing increasingly larger drains on state government budgets, contributing to 10-15 per cent of the state fiscal deficits, adversely impacting much needed investments in the social sectors of health and education. The power sector is operating with very low or no returns on equity and no contribution to future investments from internal resources. This results in inadequate investment in additional generation capacity, which is likely to further exacerbate the existing gap between power supply and demand. In 1991, IPP proposals exceeded 1,50,000 MW, while as of January 2001, just 3,500 MW of IPP power was actually operational.

Even if captive market capacity addition of 1,500-2,000 MW per year is included, a total capacity addition of not more than 6,000 MW a year over the next five years is expected. This translates into US $6 billion of investments and several million tons of additional pollutants but would still not be close enough to meet the targeted capacity increases of 1,11,500 MW by 2007. The available options ultimately are:

- Existing traditional power plants;
- Cogeneration of power;
- Competitive market for selling power;
- Renewable energy sources such as biofuels and solar that are economic and environmentally acceptable.

Demand side management (DSM), on the other hand, is a concept in which a power utility, such as a vertically integrated SEB or an unbundled distribution company, manages the demand for power among some or all its customers to meet its current or future needs. In India, DSM can be achieved through energy efficiency, which is the reduction of kWh of energy consumption or demand load management, which is the reduction of kW of power demand or the displacement of demand to off-peak times. In the former category there are programmes such as awareness generation and customer or vendor rebates for efficient equipment. The latter includes time-of-use tariffs, interruptible tariffs, direct load control, and so on. Specific type of programmes depends on the utility objective: peak clipping, load shifting, strategic conservation or strategic load growth.

Reductions in energy demand and construction at the end users' premises can free up electricity generation, transmission and distribution capacity at a fraction of costs required to provide new capacity. The cost of saved energy has been estimated to be as below as 10 per cent of the cost for added capacity for some DSM measures. The various options are:

- Use of efficient motors, energy-efficient appliances, lighting, refrigeration, heating, thermal storage;
- Load management programmes;
- Reduction of T&D losses;
- Tariff based on time-of-use;
- Revenue reconciliation;
- Use of available solar energy for heating and other applications.

## SUPPLY SIDE MANAGEMENT AND INTEGRATED RESOURCE PLANNING

First of all, the efficiency of existing power plants should be increased. Use of better quality coals, new technologies, efficient motors and modern techniques for ash handling, may increase efficiency of power plants and reduce environmental problems. Further, competitive markets for selling power are being explored. Many countries have a concept of deregulating their electric power sectors. Deregulation allows competition among generators and creates market to provide economic and higher quality of supply. The needs are achieved through transmission open access (TOA). In the open access system, the use of transmission system should be open and non-discriminatory. A recent concept, it provides better quality of service at a lower price and regulates the monopoly in the transmission system, thereby creating market conditions. Private players should also be encouraged to take interest in the business of electric power.

TOA means enabling a third party to use the transmission network which belongs fully or partially to another party for transactions under the regulatory concept. TOA may be an effective means to provide reliable and economic electric supply. For power sector reforms, TOA is a useful tool. To create competition in the electricity business it is required that there be an easy access to the electricity consumers.

In open access mode, the generators may have direct contract with the consumers. It is required that the access to transmission be regulated and efficient pricing methodologies be applied. There would be more complexity in the future expansion of the transmission system. Regulatory aspects would thus be critical.

To create a competitive electric market in the Indian scenario is a desirable feature. The present status of power supply in India is ill-conditioned. There is a need of deregulating the Indian power sector to achieve higher quality of supply at a lower cost. The power sector reform programmes have paid attention to the unbundling of three electric companies, viz., genco, transco and disco. There is very little redundancy in generation and transmission. To create open access to the transmission system, generators should be encouraged to increase capacity. As generation is less than the demand in India, a need arises to open the market for new generators. Competition in the generation sector would help to reduce the cost of supply.

At a lower scale, cogeneration and captive power are complementary solutions for supply side management. Cogeneration provides the additional demand for electric power in industries. In developing countries like India, cogeneration plants supply a significant amount of power to the industries. For better quality of supply at lower prices and to reduce risks, most industries now have their own cogeneration/captive power plants. 'A Cogeneration Facility is defined as one which simultaneously produces two or more forms of useful energy (e.g. electric power and steam electric power and shaft (mechanical) power etc.).'[1] Cogeneration is an attractive way to meet the energy requirements of industries as it is cheaper and causes less damage to the environment. Many industries in earlier days had cogeneration plants. Now again due to increasing competition in the market to offer lower price and to reduce emissions of air pollutants, they are going in for efficient use of energy. Cogeneration facilitates them to generate their own power and thermal energy from waste. The opportunities for industries having cogeneration are:

- Energy at lower price;
- Choice of fuels;
- Improved power quality;
- Utilization of waste;
- Protection of equipment from power interruptions.

Industries like sugar, fertilizers, paper, food processing and textiles have their own cogeneration plants to produce power and heat for various processes. The heat produced by the various processes can be utilized to generate power. The waste of the industries are used as primary fuels.

As far as captive power production is concerned, many industries such as steel, cement, fertilizers, aluminum, paper and sugar have captive capacity to meet significant part of their electricity needs. Captive power production fulfils about 30 per cent of the total electricity requirement of industries in India. However, this mode causes more problems due to economic returns and environment concerns.

The way to really tap cogeneration potential is to increase grid connectivity. In earlier days these cogeneration plants were not connected to the grid. The grid connectivity of power produced by a cogeneration plant depends on the following factors:

- Plant could run in synchronous or supply power radially;
- Amount of power to be supplied should be larger as compared to the line losses;
- The consumer should not be far away from the supplier;
- The cost of the laying lines and interfacing systems should be less so as to ensure profits to the cogenerator.

Technically, the cogenerator sends power at 33 kV (up to 10 MW) or 132 kV (above 10 MW) voltage through transmission lines of corresponding voltage to the grid sub-station. The cost of low voltage lines from the cogeneration plant to the grid could be borne by the cogenerator. The cost of high voltage lines could be divided between the cogenerator and purchaser. For instance, sugar mills in Uttar Pradesh sell power to UPPCL and the total cost of equipment, lines, operation and maintenance is shared between the two. The existing network may be utilized for transmitting power from the cogeneration plant to the grid on an agreement.

In regulatory terms, the commissions have made some rules to arrange these agreements:

- The plant shall maintain the equipment at the generating end and the maintenance of lines and equipments at the sub-station shall be done by the grid sub-station. The cogenerator could do any work with specific approval of the buyer or regulatory commission. Installation, operation and maintenance of interconnection should be done in accordance with Electricity Grid Code.

- The system should be designed such that if any false or abnormal operation occurs, the cogeneration plant is isolated by circuit breaker.
- The plant generating set should be synchronized with the grid. The buyer would not be responsible for any fault occurring to the plant due to failure of synchronizing or protective system.
- The power exported and imported would be measured with two sets of frequency sensitive static meters. These meters shall satisfy all the technical standards of Indian Electricity Rules and the specifications of the bureau of Indian standards.
- The transactions of power would be done as per instruction from the Load Control Centre.

## ENVIRONMENTAL ISSUES WITH COGENERATION PLANTS AND COGENERATION APPLICATION

Using waste as primary fuel, cogeneration offers some environmental benefits. It uses heat and consumes less fuel. When the fuels are burned, air pollution is the severe effect to be considered. The air pollutants are mainly nitrous-oxide ($NO_2$) and sulphur-dioxide ($SO_2$). Cogeneration reduces some of the emissions of atmospheric pollutants associated with the combustion of fossil fuels such as carbon-dioxide, a contributor to global warming, and sulphur-dioxide which results in acid rain. National Ambient Air Quality Standards (NAAQS) has set standards to meet the environmental issues for these cogeneration plants. Diesel generators produce much waste heat and are not able to utilize this heat. These generators are not environment-friendly and may cause additional air pollution. There are some special equipment required to ensure environmental regulations. With these pollution control equipment, cogeneration seems a reliable solution for industries.

Cogeneration systems can be applied for small as well as large industrial applications. Small cogeneration systems are compact and economical. Systems below 600 kW capacity produce electricity and hot water from engine waste heat which can be utilized in some other applications. The most common uses for the heat from cogeneration power plants are steam for industrial processes, and space and water heating. They can operate during peak demand to meet the additional requirement of power.

Due to the poor performance of electricity boards, frequent power cuts and high tariff, the applications of cogeneration and captive generation are increasing day by day.

To meet the increasing power demand the number of power plants is also increasing. The gas emissions from these power plants contribute to air pollution. Use of bio-fuels as a primary fuel in cogeneration power plants reduces the level of pollution and also utilizes the waste from industries. These renewable sources are available abundantly in our country and reduce the dependence on another countries for petroleum. In the IRP process, these resource options are evaluated well. For instance, bio-fuels such as ethanol and bio-diesel are becoming good supplements of petroleum fuels. Bio-fuels include alcohol, ether, ester and other bio-products such as soybean, rapeseed and vegetable. These fuels may be more economical and require less processing. Bio-products are now increasingly being seen as the bio-fuels of the twenty-first century. Biomass resources like agricultural waste, municipal solid waste, and industrial waste have been used as renewable fuels. Biomass contains high energy that has been stored through photosynthesis. That energy content remains when plants are processed into other materials such as paper and animal wastes.

Bio-fuels are produced domestically and this helps create employment opportunities and reduce our country's trade deficit. Since bio-fuels are made from agricultural waste it offers new options for the agriculture sector. It also makes us less dependent on other countries for oil and improves the energy security of our country. The environmental issue is not severe with the use of bio-fuels as they produce less harmful emissions.

## CONCLUSION

The electricity planning problem in many developing countries is more complicated than in developed countries. The challenge is not only to substitute investments in cost-effective reductions in electricity use for investment in power plants but also to obtain better efficiency of existing power plants. Issues like DSM have been discussed for years but supply side issue in power system has not been touched upon. The types of generating units used by electric utilities, cogeneration, captive plants, environmental issues are motivating forces to engage in IRP. It is estimated that IRP of supply side can increase the overall efficiency of power system by 20-30 per cent in India. Therefore, more study is needed in this area.

## NOTE

1. Ministry of Power, Resolution No. A-40/95-IPC-I, dated 06.11.1996.

# PART III

# MACRO-LEVEL ASPECTS: WHAT TO DO AS REFORMS?

# 11

## *A Benchmarking of Performance of Indian State-level Electricity Utilities in the Post-Liberalization Period*

N. BALASUBRAMANIAN

### INTRODUCTION

This study deals with benchmarking the performances of State Electricity Boards (SEB) in India. The Government of India database on Indian utilities has been used for calculating efficiency measures using the Data Envelopment Analysis (DEA). The balanced panel data for the period 1991-2001 has been used to assess the changes in the performances of Indian utilities. The study identifies the high, low and the medium performing states during the post-reform period (1991-2001) and from a set of other contextual factors of utility characteristics it identifies the reasons behind the differences in performances before suggesting improvements for future reforms in the Indian Electricity Supply Industry (ESI).

### CONTEXT

The 1991 reforms in the Indian ESI were initiated in the generation end of the industry by allowing private participation. The rationale for the liberalization at the generation end was that since generation constitutes the majority of the cost of delivered electricity, the efficiency improvements brought in by liberalization or competition in generation would lead to a decrease in the cost of electricity

*An earlier version of this paper was presented at the Hawaii International Conference on Business, June 2004.

sold. Orissa was the first state to implement a World Bank funded restructuring programme for its electricity sector. As the experience from the reforms unfolded it was found that for very many reasons reforms initiated in Orissa did not give the envisaged results, leading to a rethinking of the Indian electricity reforms. The capacity additions which were expected from private players also did not materialize leading to an exacerbation of the adverse supply-demand balance (Table 11.1). Moreover, the regional markets were not well integrated leading to the under-utilization of energy surplus in the East Zone of the country (Table 11.2). As can be seen from Fig. 11.1 India has a very high ratio of installed capacity to peak demand,

TABLE 11.1: CAPACITY ADDITIONS (IN MW)

| Five Year Plan | Target | Actual | Actual % |
|---|---|---|---|
| 1974-9 | 12,499 | 10,202 | 81.60 |
| 1979-80 (Annual Plan) | 2,813 | 1,799 | 63.90 |
| 1980-5 | 19,666 | 14,226 | 72.30 |
| 1985-90 | 22,245.25 | 21,401.64 | 96.20 |
| 1990-1 (Annual Plan) | 4,212 | 2,776.50 | 65.80 |
| 1991-2 (Annual Plan) | 3,810.80 | 3,026.50 | 79.40 |
| 1992-7 | 30,538 | 16,422 | 53.77 |
| 1997-2002 | 40,245.20 | 19,015 | 47.20 |

*Source*: Planning Commission, Government of India.

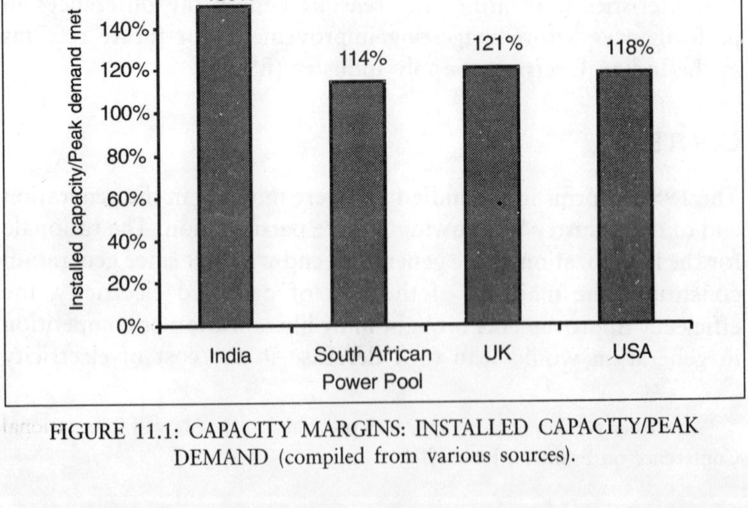


FIGURE 11.1: CAPACITY MARGINS: INSTALLED CAPACITY/PEAK DEMAND (compiled from various sources).

TABLE 11.2: REGION-WISE SHORTAGES, BACKING AND PLF

| Region | Power Shortage | Backing | PLF |
|---|---|---|---|
| North | 4.8 | 4.16 | 67.2 |
| West | 5.1 | 4.65 | 70.5 |
| South | 10.8 | 0.19 | 75.4 |
| East | –1.0 | 27.32 | 44.3 |
| North-east | 2.4 | 0 | 18.7 |

*Source*: Baijal (1999).

indicating insufficient transmission and distribution infrastructure. This is the result of decades of under-investment in transmission and distribution lines leading to the high cost of delivered electricity.

The Indian ESI is predominantly dominated by the state-level utilities called the SEBs. The SEBs own assets for distribution, generation and intra-state transmission. India has the dubious distinction of having one of the highest T&D losses for ESI in the world and exacerbating cost recoveries (Fig. 11.2). Estimates put

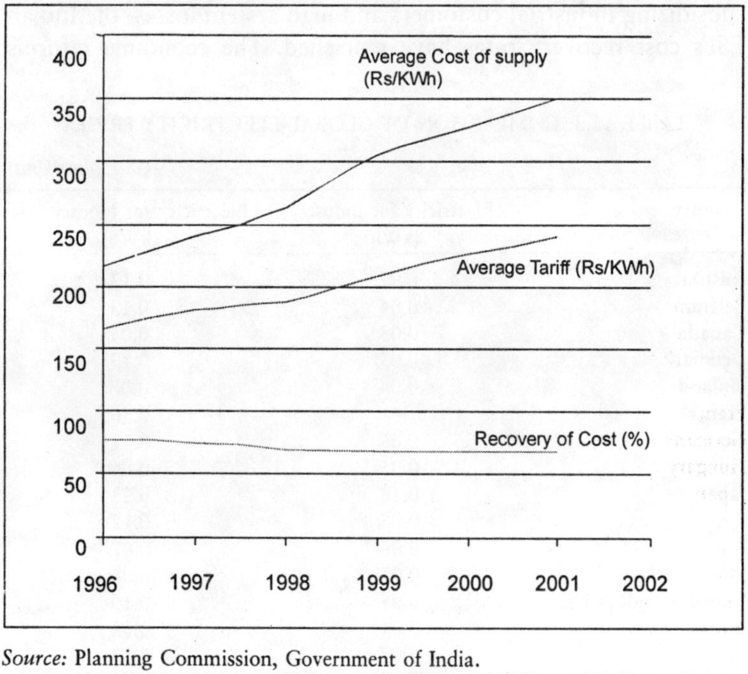


*Source:* Planning Commission, Government of India.

FIG. 11.2: COST RECOVERY IN THE INDIAN ESI.

the total system loss of energy generated (for all utilities) to be around 59 per cent owing to technical and non-technical losses and billing shortfalls. Cost recovery in the ESI has been low, leading to the financial bleeding of the SEBs. One of the main reasons for this is supply of electricity at rates below the cost to agricultural and household customers. With the burden of high electricity costs on cross-subsidizing industry, industries have gone in for small-sized captive generation. Estimates put the capacity of captive plants at 28,000 MW (Rao, 2002), which is about 28 per cent of the total installed capacity of the country. This proliferation of captive power plants though is a result of inefficiencies in the ESI, and is not an optimal solution for the problem of power. As can be seen from Table 11.3, the price of electricity for industries is amongst the highest in the world, thus affecting their competitiveness in the global markets. The share of industry in the total energy sales declined from 50 per cent in the 1970s to around 31 per cent in 2000-1 (Planning Commission, 2002). With the continuing loss of subsidizing industrial customers and high system losses, the Indian ESI's cost recovery rates have worsened. The economic reforms

TABLE 11.3: COMPARISON OF GLOBAL ELECTRICITY PRICES

*(US Dollar/Unit)*

| Country | Electricity for Industry (kWh) | Electricity for Households (kWh) |
|---|---|---|
| Austria | 0.07 | 0.13 |
| Belgium | 0.04 | 0.13 |
| Canada | 0.03 | 0.05 |
| Denmark | 0.05 | 0.19 |
| Finland | 0.04 | 0.07 |
| France | – | 0.10 |
| Germany | 0.05 | 0.12 |
| Hungary | 0.05 | 0.06 |
| Japan | 0.16 | 0.23 |
| Spain | 0.05 | 0.12 |
| Korea | 0.06 | 0.07 |
| Sweden | 0.03 | 0.08 |
| United Kingdom | 0.05 | 0.10 |
| United States | 0.04 | 0.08 |
| India | 0.07 | 0.03 |

*Source:* Rajya Sabha Starred Question No. 136, dated 27 November 2002.
*Note:* Prices are for the year 2000.

ushered in the 1980s saw some liberalization in the sector, but the more significant reforms started in the 1990s have not succeeded in dealing with these problems in the ESI. The continuing weak performance of the SEBs is a cause of serious concern and is a major impediment for successful implementation of any reforms. Hence, assessing the performance of the electricity utilities and looking for ways to improve their performances will significantly help the electricity-constrained Indian economy.

## DATA ENVELOPMENT ANALYSIS (DEA)

DEA uses a linear programming technique to fit a production or a cost frontier which is used to compare the performance of the incumbent firms in the industry. The efficient firms get to determine the frontier and other firms which are less efficient are enclosed by the frontier. The individual firms being evaluated are called the Decision Making Units (DMU) in DEA literature. DEA is used as a benchmarking tool. The frontier in DEA is formed by linear (or convex or any other specified functional forms) combinations of efficient firms. DEA can have a constant return to scale or a variable return to scale orientation. If the assignment of a functional form to the input-output relationship is not robust enough, the free disposal hull (FDH) methodology could also be used. The FDH is more restrictive than DEA in that here a firm is dominated by one single observed efficient firm while in DEA an inefficient firm is dominated by a hypothetical frontier formed by a combination of efficient firm(s). FDH is more 'empirical' than other types of DEA since it does not compute any frontier, but has the hazard of unjustifiably large number of firms being declared efficient because there is no other firm more efficient than them. Some of the strengths of DEA are that it is a non-parametric measure, hence there is no assumption about the probability distribution of the variable and the error term and has good applicative potential as a benchmarking tool. But the shortfall is that only a few firms get to decide the efficient frontier unlike in other methodologies like the stochastic frontier method or statistical methods (OLS or ML).

Farrell (1957) had proposed a piecewise-linear convex hull approach to frontier estimation, while Boles (1966) and Afriat (1972) suggested mathematical programming methods which could achieve the task, but the method did not receive attention until a paper by Charnes, Cooper and Rhodes (1978) which coined the term data envelopment analysis (DEA). They proposed a model which had an input orientation and assumed constant returns to scale (CRS). Subsequent papers have considered alternative sets of assumptions, such as Banker, Charnes and Cooper (1984) who proposed a variable returns to scale (VRS) model. DEA is the non-parametric mathematical programming approach to production frontier estimation. Based on the production frontier estimated, the technical efficiency and scale efficiency are calculated.

Productivity could simply be measured by the output produced per unit of input.

$$TT = y_0/x_0$$

where $TT$ is the productivity index and $y_0$ and $x_0$ are the output and input at time 0.

When multiple inputs and/or multiple outputs are involved, the multi-factor productivity is given by

$$\Pi = \frac{y_1/x_1}{y_0/x_0} = \frac{y_1/y_0}{x_1/x_0} = \frac{\Pi_1}{\Pi_0}$$

Where $\Pi$ is the multi-factor productivity and $y$ and $x$ are the outputs and inputs respectively for the time periods 0 and 1. For this paper the programme used was Data Envelopment Analysis (Computer) Program (DEAP) developed by Tim Coelli, Centre for Efficiency and Productivity Analysis, Dept. of Econometrics, University of New England, Armidale, NSW.

## DEA ANALYSIS

For the purpose of benchmarking the performances of the Indian electricity utilities, the following inputs and outputs were chosen:

*Inputs*

Generation Capacity (MWs)
T&D lines Circuit Kilometres
Number of Employees
Number of Distribution Transformer

*Outputs*

Agricultural Consumption (kWh)

Domestic Consumption (kWh)
High Tension Consumption (kWh)
Low and Medium Tension Consumption (kWh)
Number of Agricultural Pumpsets Energized
Traction and Miscellaneous Consumption (kWh)
Per Capita Consumption of Electricity in the State (kWh per capita)

The sources for the above data are various organizations of the Government of India which archive the state utilities' data. For the benchmarking of the state-level utilities a balanced panel data for the period 1991-2001 was used to compute the technical efficiency change with variable returns to scale (VRS) and the Malmquist TFP (Total Factor Productivity). As shown in Fig. 11.3, there is a clear separation of three performing bands of Indian electricity utilities. The first band consisting of Rajasthan, Kerala, Bihar and Andhra Pradesh show the highest change in technical efficiency over the period 1991-2001. The next band is the band of middle performers like Orissa, Maharashtra, Madhya Pradesh, Gujarat, Himachal Pradesh, Karnataka, Punjab and Tamil Nadu. The third band is the low performing states of Haryana, Assam and West Bengal.

From the analysis of the patterns, it was found that there is no correlation between the levels of reforms in the states, like the formation of regulatory commissions, unbundling and privatization of distribution, and the change in efficiencies over the period 1991-2001. In fact, the state of Bihar which shows below national average figures for measures of single factor productivity like number of employees per MWh of electricity delivered and employees per 1,000 customers has shown one of the highest changes in efficiencies. This is attributable to the fact that Bihar being a low-performing state has had the low-base effect. This refers to the fact that it is much easier to improve efficiencies when initial efficiency levels are low, than improving efficiencies when the efficiencies are already reasonably high. Of the states with high efficiency change, only Andhra Pradesh has been in the forefront of the reforms. Andhra Pradesh is an example of a well-performing state that has implemented reforms in such a way that it has actually improved the efficiencies of the utility.

All the states with low efficiency change over the period 1991-

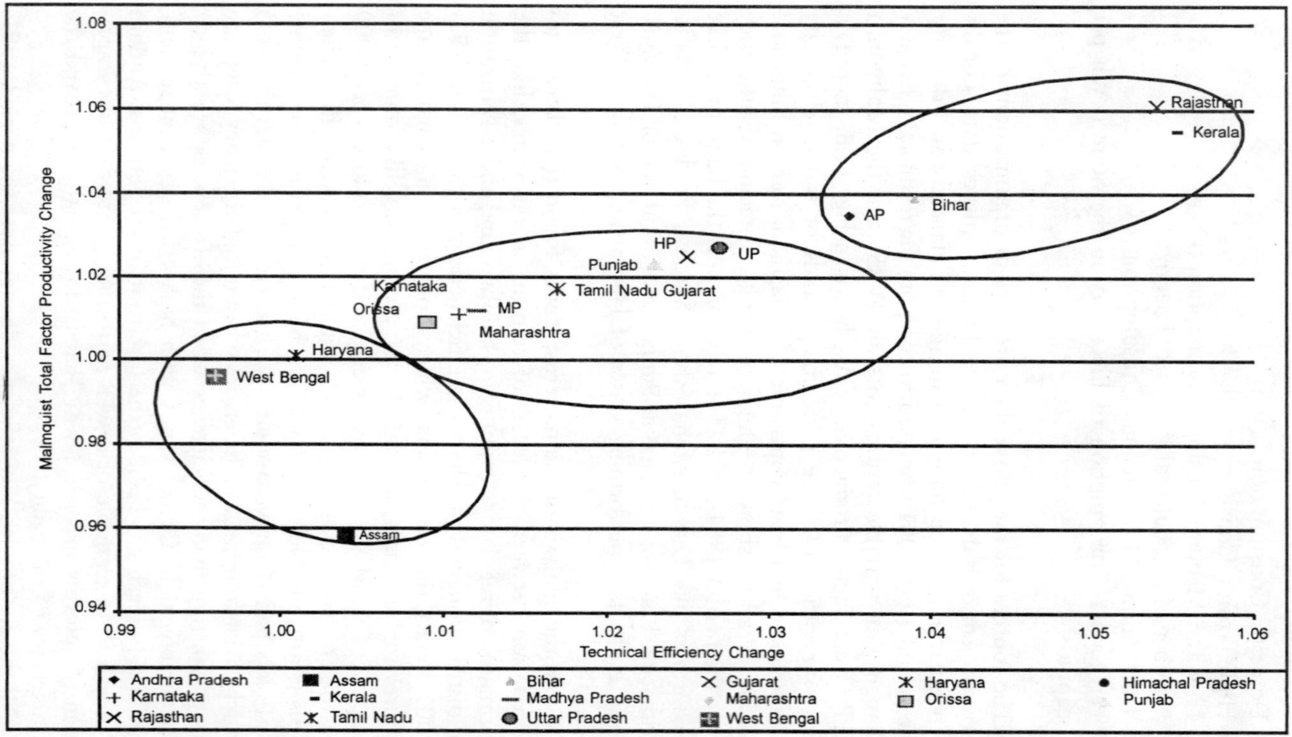


FIGURE 11.3: TECHNICAL EFFICIENCY CHANGES AND TOTAL FACTOR PRODUCTIVITY CHANGES IN INDIAN UTILITIES

2001 have low percentage of industrial consumption (Table 11.4). This points to the fact that the absence of a large industrial customer base gives poor incentive for utilities to improve their performance. This is mainly due to the fact that industrial customers are more discerning electricity consumers capable of putting forward their demands through industry associations, demand system improvements and improvements in the quality of power, while the other consuming categories like domestic and agriculture though might demand supply of power are not organized and cogent enough to push through improvements in utility performance. From the utility point of view, having low industrial load would mean lower incentive to improve efficiencies. This is so because since all the consumers except the industry and commercial are subsidized, the efficiency improvement of the utility is unlikely to result in improved revenues or better financial performance because most of the consumers are subsidized and the number of high paying industrial consumers is low. This is clearly shown when we compare the efficiency changes in the states of Haryana, Himachal Pradesh and Punjab which have similar economic environments and are contiguous. Haryana having relatively lower percentage of industrial load has efficiency changes that are lower than that of Himachal Pradesh and Punjab. The utilities whose efficiency change has been low have significantly above national average cost of electricity supply, employees per 1,000 consumers served and employee per unit of electricity delivered for the period 1991-2001. For the period of analysis all utilities were under considerable revenue shortfall and investment crunch, hence it is plausible that cost advantage could have been a significant factor driving utility efficiency changes. Since utilities with low cost of service could easily improve efficiency because they would need lower revenue and investment and hence are not financially as constrained as the utilities with high cost to service. But the findings show no such relation between cost to serve of utilities and their efficiency change. Though Assam might have adverse cost structures because of hilly terrain and low demand density, the other states having low efficiency changes do not have such extenuating circumstances.

The states with high efficiency changes show categorical advantage in neither cost of supply nor superior employee per 1,000 consumers or employee per unit of electricity sold ratio. Hence, this could imply that cost advantage is not a driver for high efficiency

TABLE 11.4: SOME CHARACTERISTICS OF THE INDIAN STATE-LEVEL UTILITIES (SEBs)

| States | Per Capita Electricity Consumption 2000 | Employees/ 1000 Customers 2001 | Employees/ MkWh 2001 | Cost of Supply 2001 | Cost of Supply 1991 | Share of Industry in Total Sales (%) 1992 | Share of Industry % in Total Sales 2001 |
|---|---|---|---|---|---|---|---|
| Andhra Pradesh | 356 | 4.6 | 2.61 | 360.7 | 89.2 | 35.0 | 24.9 |
| Assam | 72 | 18.8 | 8.80 | 589.1 | 281.2 | 27.0 | 15.0 |
| Bihar | 138 | 11.2 | 3.38 | 377.1 | 176.5 | 48.0 | 47.6 |
| Gujarat | 664 | 6.7 | 1.46 | 365.4 | 132.3 | 35.1 | 28.4 |
| Haryana | 477 | 10.6 | 4.55 | 411.9 | 115.5 | 10.6 | 18.8 |
| Himachal Pradesh | 364 | 11.6 | 4.93 | 235.4 | 118.2 | 41.7 | 37.0 |
| Karnataka | 352 | 3.5 | 5.07 | 374.6 | 92.0 | 34.6 | 21.8 |
| Kerala | 322 | 4.2 | 2.38 | 347.3 | 81.3 | 44.8 | 34.5 |
| Madhya Pradesh | 306 | 10.9 | 3.34 | 324.9 | 121.5 | 43.0 | 25.6 |
| Maharashtra | 488 | 8.6 | 2.37 | 357.5 | 124.5 | 37.8 | 38.1 |
| Orissa | 166 | 6.5 | 1.29 | 184.9 | 71.8 | 55.4 | 40.0 |
| Punjab | 800 | 15.7 | 3.65 | 285.2 | 98.8 | 34.1 | 37.8 |
| Rajasthan | 297 | 6.6 | 2.96 | 368.2 | 113.3 | 38.2 | 28.4 |
| Tamil Nadu | 552 | 6.5 | 2.57 | 309.8 | 103.3 | 41.8 | 35.6 |
| Uttar Pradesh | 155 | 6.7 | 3.74 | 383.6 | 119.5 | 26.6 | 18.0 |
| West Bengal | 187 | 9.6 | 3.79 | 376.8 | 163.7 | 29.2 | 28.2 |
| India | 308 | 7.6 | 2.53 | 350.3 | 116.8 | 35.4 | 35.3 |

*Source*: Planning Commission, Government of India.

changes in the Indian utilities and the efficiency changes are more driven by other factors like institutional effectiveness and improvement in governance and policy making in the states.

## CONCLUSION

The performances of utilities were independent of restructuring strategy, electricity reforms like unbundling, presence of an independent regulatory commission and such other sector-level reforms-related factors. Hence, the factors that could push efficiencies in the electricity sector could be more fundamental ones like strengthening of institutional set-ups at the state level, better policy making and good planning and governance in the states. If the reforms implemented are to have some effect on the efficiencies of state-owned utilities, they should be designed in such a way that utilities with diverse characteristics have some incentive to utilize the provisions of reforms to augment their efficiencies. The utilities with low industrial consumer bases did not have enough incentive to improve performances because the reforms which were implemented did not have much emphasis on the supply of electricity to agricultural, rural and domestic consumers and were industry and urban centric. Moreover the reforms focused on financial turnaround of the utilities without much concern for the overall costs and benefits of the process. The design of the future reforms process could be improved by taking into account the findings that the state-level utilities have diverse needs for incentives to improve performance and also that, rather than limiting reforms to focus on financial turnaround of the sector, having a broader perspective by emphasizing on institutional strengthening and better governance could be more effective.

## NOTE

1. Subhash Ray, 'On an Extended Decomposition of the Malmquist Productivity Index', Working Paper, University of Connecticut, Storrs, CT 06269-1063.

## REFERENCES

Afriat, S.N., 'Efficiency Estimation of Production Functions', *International Economic Review*, vol. 13, no. 2, 1972, pp. 568-98.

Baijal, Pradip, 'Restructuring Power Sector in India, A Base Paper', *Economic and Political Weekly*, 25 September 1999.

Banker, R.D., A. Charnes, W. Cooper, 'Some Models for Estimating Technical and Scale Inefficiencies in Data Envelopment Analysis', *Management Science*, 30:9, 1984.

Boles, J.N., 'Efficiency Squared—Efficiency Computation of Efficiency Indexes', Proceedings of the 39th Annual Meeting of the Western Farm Economics Association, 1966, pp. 137-42.

Charnes, A., W. Cooper, E. Rhodes, 'Evaluating Program and Managerial Efficiency: An Application of Data Envelopment Analysis to Program Follow Through', *Management Science*, 27:6, 1981.

———, 'Measuring the Efficiency of Decision Making Units', *European Journal of Operational Research*, 2:6, 1978.

Caves, Dougloss W., Laurits R. Christensen and Erwin W. Diewert, 'The Economic Theory of Index Numbers and the Measurement of Input, Output and Productivity', *Econometrica*, 50(6), 1982.

Fare, Rolf, Showna Grosskopf, B. Lindgren and P. Roos, 'Productivity in Swedish Pharmacies: 1980-1989 A Nonparamatric Approach', *Journal of Productivity Analysis*, 3, 1992.

Fare, Rolf, Showna Grosskopf, M. Norris and Z. Zhang, 'Productivity Growth, Technical Progress and Efficiency Change in Industrialized Countries', *American Economic Review*, 84, 1994, pp. 66-83.

Farrell, M.J. 'The Measurement of Productive Efficiency', *Journal of the Royal Statistical Society*, Series General, 120(3), 1957.

Rao, S.L., 'The Political Economy of Power', *Economic and Political Weekly*, 17 August 2002.

Ray, Subhash and Evagelia Desli, 'Productivity Growth, Technical Progress and Efficiency Change in Industrialized Countries: Comment', *American Economic Review*, 87(5), December 1997, pp. 1033-9.

Ray, Subhash and Kankana Muckerjee, 'Inter-State Differences in Productivity Growth in US, Manufacturing: A Nonparametric Analysis of Census of Manufacturers Data, Working Paper, University of Connecticut, Storrs CT 06269-1063.

Planning Commission, *Annual Report 2001-2002 on The Working of State Electricity Boards and Electricity Departments*, Government of India, 2002.

Shephard, R., *Cost and Production Functions*, Princeton: Princeton University Press, 1953.

# 12

## *Generation Choices: A Retrospective Account of the Ninth Plan*

JOËL RUET

This paper focuses on the profitability of investments in the power sector in India, with a micro-economic concern followed by macro-extrapolation. We compare the profitability of various investment strategies that are alternatives to simple expansion of the generation capacity. In a system where maintenance has long been neglected, these measures also turn out to be cheaper to capacity addition, and should be favoured in a period of scarcity of funds. These measures include improvement in the use of plants and reduction of technical and non-technical losses. Though profitable, these are not implemented under the present public system in India.

We then briefly link up these findings with some of our earlier works to suggest that these inefficiencies arise because the State Electricity Boards (SEB) have long relied on technical and administrative procedures that were designed during a period of expansion, and not on more efficiently decentralized and information-based management as should be the case in the present maturity phase.

### INTRODUCTION

Several studies deal with the investments required by the Indian power sector to bridge the gap in demand and supply. Most of these are based on the business-as-usual scenario and merely extrapolate the current trends, without looking much at efficiency improvement. These projections on future demand, for instance in Mohan (1997) and Shukla (2000) show large upcoming equipment needs, regardless of the considered scenario. The business-as-usual hypotheses show that India lacks money for its power system development. This approach is in line with the official position on reforms. The hypotheses also discusses the inefficiencies in the operation of SEBs. The Government of India (1998a and 1998b) still contends that 'the most serious problem facing the power sector is the paucity of investment funds'. It argues (Government of India 1998b, 2004) that additional resources for system upgradation would need private sector participation in generation, transmission and distribution. Significantly, the need for reforms seems to emanate from these concerns, though it is recognized that to attract private capital to the industry, structural reforms would be required (Government of India, 2004; Ruet, 2002 & 2005). This issue is hence central in the debate on power reforms. However, these macro-studies are based on more or less conservative hypotheses, at least as far as the managerial efficiency in the SEBs is concerned.

Another set of studies, essentially led by the World Bank (1996, 1997, 1999), for the reforms in the states of Orissa, Haryana and Andhra Pradesh has attempted at quantifying the need for structural reforms in operation and management, and restructuring of generation. It provides interesting accounts of possibilities for thinking beyond the frame of increasing public expenditure in expanding the generation capacity, sifting through the many possibilities of technical-cum-managerial reorganization of the power sector.

We want to complement these studies, in three ways. First, these studies don't bring out clearly the economic profits that can come out of these particular measures, in terms of internal profitability, for instance. We propose this calculation here, based on several hypotheses and proxys for some variables, but by stating out clearly the methodology. Second, the extrapolation of these results to an all-India level, and not limiting them to state-wise analysis brings an idea of the scope for such reforms. Third, some experiments on better maintenance and rehabilitation of assets have been attempted over the last five years by some public utilities. We will not give a systematic account of them here, but will indicate how they could be incorporated in order to finetune some evaluations of the investment needs to reform the power sector in India.

Investments profitability is more precisely examined here, according to different technical and commercial choices. We use 1996-7 as well as 2001-2 figures as a reference, in order to show the trend over five years. The year 2001-2 is chosen for reasons of harmonization in figure availability (the last year for which all figures are available). We look at short-run profitability, not making assumptions about future growth, nor the long-run marginal development cost. We first present the data then we present the various technical options available, for which we propose a calculation methodology to address fixed and variable costs and annual returns of several technical strategies at given tariffs. The possibilities we examine are Plant Load Factor (PLF) improvement and reduction of technical losses. For sake of reference, we also compute the costs and incomes of a commercial policy that tackles non-technical losses. Several costs and incomes depend on the level of tariffs. Hence, we investigate the impact of tariff increase (based on present costs) on the outcome of each strategy. These issues are dealt with separately. Then, we compare all strategies with classical thermal capacity addition. Of course, in each case the corresponding network investment to evacuate the power is taken into account, as well as the incidence on operation costs.

In the section on techno-economic discussion, we discuss the profitability of various strategies in both tariff contexts, and analyse the margins on actual tariff. We argue that even under the current subsidized tariffs—on which SEBs argue that they lack money for investment and operation—several clearly profitable strategies exist, that are not enforced by SEBs.

We conclude by suggesting how the internal organization of SEBs precludes them from taking profitable decisions.

## MAIN DATA[1]

An important point is that the costs that we compute are those faced by the SEBs and their shareholders, the state governments (direct costs). We do not here focus on indirect revenues for the state generated from taxes, nor is attention paid to additional growth of consumers, nor to any ousting effect. Studies focussing on Orissa, Haryana and Andhra Pradesh by the World Bank (1996, 1997, 1999) show that with such kinds of reforms, the global economic profitability goes along with the SEBs' financial viability. The point here is only to show that such actions are profitable, and to add evidence to the fact that SEBs are not taking economically sound decisions.

In order to compare different strategies, we will look at them in isolation. However, since the various strategies do not present the same scope and margins in terms of GW of capacity to be added, we will also focus on sets of combined strategies. We chose something linked to what in the eye of the public is exemplary of the sector's failure: the power cuts that arise because of lack of energy. We focus on strategies that allow bridging the energy gap, that is, the yearly gap between demand and supply—hence the title of this chapter.

As per the Planning Commission (2002), the energy gap is 7.5 per cent of the demand (it was 11.5 per cent in 1997). From interviews in NTPC, the development of thermal plants is considered as costing Rs. 4 crore[2] per MW. As underlined by the Planning Commission (1997), a 1:1 ratio has to be taken into account to develop the associated evacuation networks, hence an additional Rs. 4 crore. In our calculations, new plants that are developed to bridge the gap are supposed to have a good PLF. We have taken a PLF of 68.5 per cent for the calculations of 1996-7, which was the norm required from the independent power producers (compared to the actual Indian average of 64 per cent and SEBs' average of 60.1 per cent in 1996-7 (Planning Commission, 1997). We have taken it to be 70 per cent for the calculations relative to 2001-2, taking into account the relative progress accomplished in-between (with an actual Indian average of 69.9 per cent and SEBs' average of 67 per cent, and 74.3 for the Central-level utilities in 2001-2 (Planning Commission, 2002).

Variable cost for fuel is taken as equal to 90.90 paisa per kWh in 2001-2 and 71 paisa per kWh in 1996-7.[3] Variable cost for wages, operation and maintenance, and administration is 36.70 paisa per kWh for generation, transmission, and distribution (source: Calculations by the author, based on Planning Commission data; where it represents 15.30 per cent of the total costs; the figure can be similarly calculated for 1996-7; we get 35.7).

Energy losses are another key parameter. From the examples of Orissa, Haryana, Andhra Pradesh, and Delhi, it may be assumed that the total figure of 27.80 per cent (22 per cent in 1996-7) officially reported by SEBs to the Planning Commission is fake. Orissa, Haryana and Andhra Pradesh used to declare 24 per cent, 25 per cent and 19 per cent respectively as level of losses. The setting up of a more transparent reporting system following the reforms revealed these figures to be as high as 46 per cent, 34 per cent and 33 per cent for Orissa, Haryana and Andhra Pradesh, respectively. This figure could probably be 35-40 per cent in Haryana, according to interviewed consultants. Though, Punjab is reporting 18 per cent loss according to interviews held, it could be at least 10 per cent more. Delhi used to declare 23 per cent loss, but this has officially gone up to 43 per cent and has since then peaked at 54 per cent.[4] Therefore, it is more realistic to consider the total losses to be far higher than the official figures. Indeed, till today, some States are not fully transparent or not fully metered (for instance, only 88 per cent of consumers are metered in Maharashtra, 85 per cent in Punjab, 63 per cent in Madhya Pradesh, 50 per cent in Bihar; source: Ministry of Power, 2003). World Bank (1998) indicates that total losses are in the range of 40 per cent to 50 per cent for entire India. Since, in absence of 100 per cent metering, one can hardly distinguish between technical and non-technical losses, a conservative assumption is to consider that non-technical losses cover at least half of them. Indeed, according to interviewed people and a few energy audits in the country (Andhra Pradesh in 1996, Delhi since 2002 onwards), non-technical losses are certainly higher than technical ones. We will therefore consider that both technical and non-technical losses are at least 22 per cent (equal to the official figure for technical losses), the sum being 44 per cent, a figure consistent with the World Bank (1998).

Finally, the last figure we need to devise is the extra capacity needed to cover the gap. In 1996-7, the gross generation was 394 TWh (515 in 2001-2). With a deficit of 11.50 per cent and a PLF of 68.50 for new plants, additional capacity of 7.60 GW was necessary (figures for 2001-2 are 7.5 per cent, PLF taken at 70, and thus a needed capacity of 6.30 GW only).

Let us summarize these figures in Table 12.1.

## TECHNICAL STRATEGIES

The Indian electric system has developed essentially through addition of new capacity, and by neglecting maintenance (cf. Planning Commission, 1997 to 2002; Government of India, 2004; World Bank 1996, 1997 and 1999). This paper argues that due to inadequate attention to other techniques in the past, there now exists large profitable margins of gains to meet the growing demand.

TABLE 12.1: MAIN DATA RELATIVE TO GENERATION 1997 AND 2002

| Parameter | 1997 | 2002 |
|---|---|---|
| Gross generation (TWh) | 394 | 515 |
| Energy gap (%) | 11.5 | 7.5 |
| Actual PLF for SEBs (%) | 60.1 | 67.0 |
| Normative PLF (%) | 68.5 | 70.0 |
| Extra capacity needed to bridge the gap at normative PLF (GW) | 7.6 | 6.3 |
| Recovery of costs through tariffs (%) | 76 | 68.6 |
| Non-technical losses (%) | 22 | 22 |
| Technical losses (%) | >10 | >10 |
| Variable cost for fuel and energy purchase (mix at SEB sourcing ratio) (paise/kWh) | 71 | 90.9 |
| Variable cost on wages and Operation and Maintenance (paise/kWh) | 35.7 | 36.7 |
| Development cost (crores/MW) | 4 | 4 |

*Source*: Planning Commission, NTPC, Ministry of Power, personal hypothesis, interviews.

Let us review some alternative strategies.

## STRATEGY 1: BRIDGING THE GAP WITH NEW PLANTS

Let us explain step by step how we have carried out our calculations. At Rs. 4 crore/MW, the need to develop 7.60 GW (6.30 in 2002) leads to an investment cost of Rs. 30,200 crore in generation for the 1996-7 situation (Rs. 25,200 with 2001-2 figures), to be completed by an identical amount in T&D network. The total investment required for the power sector is then Rs. 60,400 crore, with generation and associated network at 1:1 ratio (Rs. 50,400 in 2001-2). The annual variable cost of generated energy is calculated by adding generation and network costs (Rs. 71 + Rs. 35.7 for 1996-7 and Rs. 90.90 + Rs. 36.70 for 2001-2); the network variable costs are necessitated because networks have also to be developed in this strategy to evacuate energy. The calculation gives Rs. 4,850 crore annually (Rs. 4,930 for 2001-2).[5] Conversely, with a 76 per cent recovery on cost from tariffs for 1996-7 and 68.6 per cent for 2001-2 (Planning Commission, 1997 and 2002),[6] the revenue is Rs. 7,150 crore, as seen for 1996-7 and Rs. 9,270 crore for 2001-2. The net annual revenue for SEBs is

Rs. 7,150 – Rs. 4,850 = Rs. 2,300 crore in 1996-7
and Rs. 4,340 crore for 2001-2.

STRATEGY 2

Bridging the gap with preliminary PLF improvement of the existing plants up to 68,50 per cent (70 per cent as far as 2001-2 capacities are concerned).

Improving the PLF can ensure better energy availability. This allows lowering the need for extra capacity. We calculate the capacity-equivalent that the PLF represents. PLF improvement of the existing plants from 64 per cent to 68.50 per cent (1996-7 figures) would have resulted in an improvement margin of 4.4 GW of equivalent capacity. Similarly, improvement from 67 per cent (SEB figures) up to 70 per cent (2001-2 figures) results in an improvement margin respectively of 2.1 GW of equivalent capacity.[7] We evaluate the cost of renovation and improvement to be Rs. 2,800 and Rs. 1,340 crore respectively.[8] With these improvements, the power required to bridge the gap is brought down to 3.2 and 4.2 GW respectively. With calculations similar to the previous case, this leads to an additional investment cost of Rs. 25,200 crore (1997) and Rs. 33,760 crore (2002) in generation plus network (the 1:1 ratio hypothesis for generation and network is again taken in calculations for arriving at these 3.2 and 4.2 GW). But no network investments are considered for the additional capacity coming from PLF 'equivalent-capacity' improvement (4.4 and 2.1 GW, respectively). Indeed, in first approximation, the existing network was designed to yield peak capacity.[9] The total investment is hence Rs. 28,000 crore for 1996-7, and Rs. 35,100 crore for 2001-2.

For 1997, since the PLF improvement does not include network expenses, only (71 + 9.7) paise per kWh is charged as variable cost attached to the 4.4 GW; the other part is treated as before. The total annual variable cost of generated energy is then Rs. 4,150 crore, still with a revenue of Rs. 7,150 crore. The net revenue is then Rs. 3,000 crore for 1996-7. Similar calculation for 2001-2 gives an annual variable generation cost of Rs. 4,580 crore and an annual revenue of Rs. 9,270 crore.

STRATEGY 3: BRIDGING THE GAP WITH PRELIMINARY TECHNICAL LOSSES IMPROVEMENT

Let us examine how much it costs to reduce technical losses by 10 per cent. This evaluation is most arguable; therefore, we have conducted several estimations to cross-check, based on different states and confirmed through implementation in small zones. These estimations converge to the same result. We made an evaluation based on information available for Haryana, Orissa, and Andhra Pradesh, as well as from the experiments of the private Noida Power Company.[10]

EVALUATION 1

Haryana's reform implied a requirement, in phases 1 and 2 (immediate action and structural improvement action), of 277 million US Dollars, i.e. Rs. 1,160 crore (World Bank, 1997). If we consider that the investment in phase 2 is partly designed to prepare phase 3, more specifically designed to meet the growth, then we can consider the sole rehabilitation figure of Rs. 930 crore over six years (World Bank, 1997). Since an improvement of losses by 1.4 per cent per year on average was planned, this investment covers a gap of 8.4 per cent. Reducing losses by 10 per cent according to this estimation would require Rs. 1,100 crore. For the state of Haryana the extrapolation of these figures, based on proportionality of consumption to the entire country will amount to an estimation of Rs. 35,000 crore.[11]

EVALUATION 2

Andhra Pradesh power reform involves 103, 70, and 32 million US Dollars respectively for transmission system improvement (voltage improvement, overloading and losses reduction), distribution system, and technical assistance (World Bank, 1999). This investment targets, over four years, an expected improvement of 0.5 per cent per year. An extrapolation of these for the entire country stands at Rs. 33,000 crore.[12]

EVALUATION 3

In the same way, in Orissa a Rs. 1,100 crore project was explicitly targeting a 12 per cent reduction over five years (World Bank, 1996), leading to 10 per cent, or to the figure of Rs. 35,000 crore (for the entire country).[13]

Extrapolating World Bank state-level studies to the entire country gives robust results. Let us however cross-check from another source.

EVALUATION 4

Let us check if one practical experience in the small Noida Power Company[14] gives results comparable to evaluations 1, 2 and 3. Actual improvements have brought down losses from 8 per cent to 7 per cent between 1995-6 and 1997-8. From the financial year 1994-5, in which investment in NPC was mainly dedicated to follow the demand growth (NPC, 1995), one can calculate the marginal development cost in terms of crores per added MWh of demand. Then, from total investment in networks for the two concerned years, the part concerning development *per se* can be separated from that concerning losses reduction. Once the development part (development cost multiplied by the increase in demand) is ascertained, the rest can be attributed to losses improvement actions. This gives an amount of Rs. 275 lakh to achieve loss improvement of 1 per cent.[15] Linear extrapolation of these figures to entire India gives a total of Rs. 8,500 crore for a 1 per cent reduction. However, from this figure, in order to evaluate for a 10 per cent improvement, calculation must be made more carefully, since a 1 per cent improvement from 8 per cent is harder than a 1 per cent improvement in a situation where technical losses are 22 per cent. The ratio between the two levels of losses has to be taken into account and the quadratic form of technical losses has to be considered. We get by this method Rs. 31,000 crore.[16] Hence, for a 10 per cent reduction in technical losses, the figure of Rs. 35,000 crore still stands.[17]

Now let us evolve a full cost-benefit analysis of such a strategy, as we did for the previous ones. A 10 per cent improvement applied to the 1996-7 figures will result in an additional 6.6 GW of equivalent capacity of power, leaving only 1 GW necessary in the form of new plants and associated networks (the latter part costing Rs. 8,000 crore). Investments necessary for network improvement, metering, information system upgrading are difficult to assert. However, from the four previous converging estimates, let us take the figure of Rs. 35,000 crore. The total investment is then Rs. 35,000 + Rs. 8,000 = Rs. 43,000 crore for 1996-7. Similarly, for 2001-2, the gap being of 7.5 per cent, a 10 per cent reduction would more than cover it. Let us then focus on a 5 per cent improvement only, which can be considered as being even more feasible (according to bids given during the privatization of distribution in Delhi, this is reachable within three years). Evaluating the investment cost in 2002 using the 1997 figures, we will consider it as proportional to the energy it allows to save, as between 1997 and 2002 figures, the network has grown homothetically to the capacity at the first order.[18] Taking into account the load growth, we get an investment of Rs. 22,900 crore with 2002 figures. This investment provides an additional 4.2 GW equivalent capacity, hence 2.1 balance need being bridged through new plants (plus network), with a respective associated costs of Rs. 16,800 crore. Total investment for 2002 is then Rs. 39,700 crore.

No variable costs remain for the part associated with the improvement in technical losses (energy was already generated in the reference situation). Only Rs. 650 crore of variable costs remain, still associated with a revenue of Rs. 7,150 crore, leading to a net revenue of Rs. 6,500 for 1996-7. For 2001-2, variable cost is similarly Rs. 1,640 crore, with a revenue of Rs. 9,270, therefore a net revenue of Rs. 7,630 crore.

It may be noted that combining improvement in both PLF and technical losses solves the problem without any additional capacity requirement.[19]

## TACKLING NON-TECHNICAL LOSSES AND TARIFFS ISSUE

In this section, we examine the direct cash-flow effect (for the power sector) of suppressing non-technical losses. We take examples of states which have made an audit, or for which losses have been made transparent, namely Orissa, Haryana, Andhra Pradesh and Delhi.

Regarding non-technical losses suppression, experience shows that these losses occur in all categories of users. However, since the actual allocation between these is not easily available due to lack of 100 per cent metering, an assumption has to be made. We assume that they do not appear in the industrial sector, and that they are shared in proportion to their consumption by other sectors. However arguable, this assumption provides a mark-up, and is chosen here to minimize the profitability of the suppression of non-technical losses. Indeed, given the cross-subsidization of the power sector, this assumption lowers the potential incomes of such a measure. Hence, the actual profitability will be higher than the one computed here and we remain on the safe line. Since the article proves that under these hypotheses, the measure is profitable, the assumption is consistent with result robustness.

Under 1996-7 circumstances and at then tariffs, the suppressing of non-technical losses would have yielded Rs. 5,000 crore annually (the 22 per cent figure being billed at average tariff excepting industrial users, after the figure of Planning Commission, 1997; with this methodology, the additional income is ventilated as: Rs. 2,350 to come from domestic, Rs. 1,610 from commercial, and Rs. 1,000 from agriculture, if one takes into account the different tariffs category-wise, as provided by the Planning Commission report). This of course implies an organizational cost, since metering and training programmes are necessary. The amount may be evaluated from consulting expenses in several reforming states as being inferior to an investment of Rs. 3,700 crore for all of India,[20] i.e. an annuity of Rs. 600 crore. No additional variable cost has to be considered, since these losses were already actually generated and consumed in the reference scenario. This action would have been extremely profitable. Similarly, for 2001-2 figures, the extra revenue is of Rs. 12,550 crore, with a break-up of Rs. 6,950 crore from domestic, Rs. 3,630 crore from commercial and Rs. 1,980 crore from agricultural consumers. This formidable increase latter is explained by the fact that both tariffs and consumption have risen in the five years. Let us for illustration show what a targeted tariff increase occurring right from 1996-7 would have brought.

## The Impact of Tariffs Increase

The decision to increase tariffs is different from other strategies and does not involve any investment. It is, therefore, obvious that it will generate an income. Our aim then is only to assess the level of the stake by analysing what becomes the profitability of precedent strategies in a context of prior tariff increase.

### INCREASE OF TARIFFS TO REACH THE PRESENT AVERAGE COST OF SUPPLY

If non-technical losses remain at current level (22 per cent), and energy sold is paid at average cost (208 paise/kWh, according to the Planning Commission, 1997), an income of Rs. 10,050 crore accrues. Several tariff adjustments were available to reach this result.

For instance, had there been political will not to charge industrial consumers more, the average tariff for other categories would have been 173 paise/kWh (instead of 97 paise/kWh for an average cost of supply of 208 paise in 1996-7, and actual average recovery of 158 paise, as given by the Planning Commission, 1997).

Let us now assess the cash flows under a regime of increased tariffs. For each technical strategy, the investments and variable costs remain the same, but their annual revenue (that is, only on the additional 11 per cent given by the bridging of the energy gap) if charged at 1997 tariffs, is Rs. 7,150 crore, and with increased tariffs, is up to Rs. 9,400 crore.

As far as non-technical losses are concerned, if combined with a prior increase in tariffs, their suppression provides a yearly income of Rs. 12,850 crore,[21] instead of the previous figure of Rs. 5,000 crore. Of course, the cost of investment has still to be made. We will see, interestingly, that an equivalent rise in tariffs has actually occurred over the last five years, and we will analyse how it has impacted the profitability of the strategies we have considered.

## TECHNO-ECONOMIC DISCUSSION

Let us now discuss the profitability of each action.

Investment (*I*), variable costs (*VC*), annual revenue (*R*) and cash flow (*CF*) are shown in Table 12.2, where $CF = R - VC$. They are analysed both with current tariffs, and also in the context of the increase in prior tariffs at average cost as far as 1996-7 is concerned (208 paise/kWh).[22] The internal rate of profitability is calculated with the hypothesis of a long duration of life of assets (valid for the power sector).

### BRIDGING THE GAP

Clearly, building new plants within the present average tariffs is the least profitable, and even not profitable if one is looking for a 16 per cent rate of return, which is now the granted level for independent producers and central government undertakings as per Electricity Act 2003. This represents the long-term tension of the power sector in India—the combination of political fixation of tariffs (or the inertia linked to that even in a new context of regulatory commissions) and inertia in cost-ineffective investment—leading to

TABLE 12.2: PROFITABILITY OF MIXED STRATEGIES TO BRIDGE THE ENERGY GAP

(11% gap for 1997, 7.5% for 2002)

| Strategies | Investments and cash flow (crores) | | | Internal rate of return (%) | | |
|---|---|---|---|---|---|---|
| | 1997 at fixed tariffs | 1997 if tariffs increase | 2002 at current tariffs | 1997 at fixed tariffs | 1997 if tariffs increase | 2002 |
| New plants by 7.6 GW (6.3 GW) | I: 60,000<br>VC: 4,850<br>R: 7,150<br>CF: 2,300 | I: 60,000<br>VC: 4,850<br>R: 9,400<br>CF: 4,550 | I: 50,400<br>VC: 4,930<br>R: 9,270<br>CF: 4,340 | 4 | 7.5 | 8.6 |
| New plants by 3.2 GW (4.2 GW) and PLF increase up to 68.5%, i.e. 4.2 GW (PLF to 70%, i.e. 2.1 GW) | I: 28,000<br>VC: 4,150<br>R: 7,150<br>CF: 3,000 | I: 28,000<br>VC: 4,150<br>R: 9,400<br>CF: 5,250 | I: 35,100<br>VC: 4,580<br>R: 9,270<br>CF: 4,690 | 10.5 | 18.5 | 13.4 |
| New plants by 1 GW (2.1 GW) after technical losses improvement -10%, i.e. 6.6 GW (-5%, i.e. 4.2 GW) | I: 43,000<br>VC: 650<br>R: 7,150<br>CF: 6,500 | I: 43,000<br>VC: 650<br>R: 9,400<br>CF: 8,750 | I: 39,700<br>VC: 1,640<br>R: 9,270<br>CF: 7,630 | 15.1 | 20.3 | 19.2 |
| Non-technical losses elimination –22% | I: 3,700<br>VC: 0<br>R: 5,000<br>CF: 5,000 | I: 3,700<br>VC: 0<br>R: 12,850<br>CF: 12,850 | I: 3,700<br>VC: 0<br>R: 12,550<br>CF: 12,550 | 135 | 347 | 339 |

*Note*: I = investment; VC = variable costs, R = annual revenue, and CF = cash flow.

the financial incumbency of the sector. However, we can clearly see two things:

- The effect of tariff increase since 1997 has contributed to increasing the profitability of the sector.
- More important, some strategies, like PLF improvement and losses reduction, clearly show higher profitability. Even if they have been marginally tapped since 1997, they still offer good margins for developing at cheap cost an interesting capacity-equivalent.

The strategy based on PLF improvement completed by new plants shows a net revenue that is reasonably close to allowing a 16 per cent profitability requirement (if collection rate improves, 16 per cent would be reached). PLF improvement of existing plants is less expensive and leads to the same result (partly because PLF improvement itself is less costly, and partly because this happens on existing plants, i.e. with already existing evacuating networks that are dimensioned for the annual peak, regardless of the rate of utilization.

A strategy based on the reduction of technical losses, then complemented at the margin along with the addition of new plants gives an important cash flow. The fact that it was 15 per cent in 1997 clearly shows that the situation has deteriorated, thereby allowing higher potential margins of progress today.

Hence, to bridge the present energy gap, or to match the increase in demand within the context of the remaining gap, the improvement of PLF and the reduction of technical losses prove to be more profitable strategies in relative terms. Table 12.2 shows then that improvements in operation are to be preferred to new investments. It is a fact that investment in new plants has not only the highest investment cost on annuity, but also the highest variable costs. Conversely improvements in the PLF or reduction of technical losses imply lower investment and an increased cash flow. Unfortunately, these policies are currently not adopted enough by SEBs. If one looks at the situation as it was in 1997, tariff increase seemed to be politically difficult, but interesting and politically neutral technical strategies were available. Part of them have been tapped, namely PLF increase, but in a very limited manner (the average value for SEBs is still very low compared to the value for Central utilities).

Losses reduction has been quite untapped. This shows clearly from the fact that, after five years, the margins remain (both in terms of capacity-equivalent and financial profitability). Paradoxically, the direct cash flows have followed the trend that a tariff increase would have given. The 'politically difficult' measures have been undertaken (under the sheer compulsion of equivalent cost increase), while technical solutions have remained untapped. Worse, techniques and costs are actually linked: had the technical measures been applied right from 1997, they would have comparatively lowered the average cost per unit. We are in a strange scenario when the easily available margins have remained unused, while cost inefficiency has left the politicians with no choice but to do what they wanted the least.

The reduction of non-technical losses is obviously profitable, but has to be distinguished from the technical strategies, as it does not physically help to bridge the energy gap. While it is not a productive measure, it can nevertheless have redistributive benefits by giving a margin to finance the other actions (the beneficiaries being the power companies), reducing subsidies (beneficiaries are the taxpayers), helping smooth the tariff increase (beneficiaries are the consumers), or combinations of the above.

### INTERNAL PROFITABILITY OF PLF IMPROVEMENT AND TECHNICAL LOSSES REDUCTION

We have conducted the analysis till now in the form of a potentially policy-oriented goal: bridging the gap. The above mixed strategies however show neither the specific profitability of PLF improvement, nor of the technical losses reduction taken separately. Table 12.3, briefly details them in isolation for a 1 GW-equivalent increase.

Although the reduction of technical losses in itself would mean a higher investment than that of PLF improvement, the associated variable costs are far lower. This is due to the fact that in the reference scenario, they were produced anyway. Even under the present tariffs, PLF increase is very profitable, although on low-level investment, so that finally the amount concerned is not very high. In other words, the main improvements are to come from the SEBs, both on the network side and on their generation operation (they are the utilities that have the lowest PLF).

Practically, reduction of technical losses is expected to take more time than PLF increase and our calculations could be easily refined

TABLE 12.3: ISOLATED STRATEGIES

| Strategies | Investments and cash flow (crores) | | | Internal rate of return (%) | | |
|---|---|---|---|---|---|---|
| | 1997 at fixed tariffs | 1997 after tariffs increase | 2002 | 1997 at fixed tariffs | 1997 after hypothetical tariff increase | 2002 |
| PLF increase (equivalent 1 GW) | I: 620<br>VC: 480<br>R: 920<br>CF: 440 | I: 620<br>VC: 480<br>R: 1,210<br>CF: 730 | I: 640<br>VC: 730<br>R: 1,470<br>CF: 740 | 71 | 117 | 116 |
| Technical losses improvement (equivalent 1 GW) | I: 5,110<br>VC: 0<br>R: 920<br>CF: 920 | I: 5,110<br>VC: 0<br>R: 1,210<br>CF: 1,210 | I: 5,450<br>VC: 0<br>R: 1,470<br>CF: 1,470 | 18 | 24 | 27 |

*Note*: I = investment; VC = variable costs, R = annual revenue, and CF = cash flow.

by putting a time frame of 1.5 per cent reduction per year. However, the general conclusion remains: to achieve the same result, favouring these two techniques at the highest speed possible is the best strategy. The results are also robust, in spite of the fact that analysis does not explicitly integrate the increase of demand. Of course, in such a case, the benefits of fighting against technical losses is less visible, as growth mechanically leads to their increase. But, with a business-as-usual scenario as reference, the profitability remains, as the action against technical losses will still counterbalance a rise in losses.

As far as comparison between 1997 and 2002 is concerned, we again find the same striking paradox: the State has done the unpopular work, without the technicalities of the game being addressed. Or is it internal resistance to change illustrated by this very fact? This is what we briefly discuss now.

## ELEMENTS OF ORGANIZATIONAL DISCUSSION

The fact that the above profitable actions are not undertaken in SEBs is puzzling. A case study based on the Haryana SEB may be useful in understanding how the internal organization of the SEBs itself leads to inefficient decisions. This example is consistent with a whole system, taking place in every SEB in India (for further details, see Ruet, 2001 and 2005).

The annual rate of damage of distribution transformers is abnormally high, reaching more than 30 to 35 per cent in some sub-divisions, according to field interviews. This situation is a techno-economic paradox. Indeed, from experience derived from World Bank loans, it appears that investments to increase the number of transformers by less than 10 per cent brings down the rate of damage to 20 per cent, i.e. at least an improvement of one-third, as field inquiries in Haryana show. Such investments are highly profitable but are not made. Paucity of funds is the common answer to that argument. We argue that the budgetary system itself precludes generating additional income.

Let us look at what happens when a transformer is damaged in a sub-division. It is sent to the store for replacement or repairs. Repairs are done in the workshop and returned to the store which sends it back to the sub-division to which it belongs. There is also an administrative circuit where the responsibility for damages are checked by a hierarchy on a sampling basis. The head office actually monitors the transformer's repair through a constraint on its rate of damage.

Orders for replacement are issued at the sub-division level when a damage occurs. Costs are compiled at the headquarters on an average basis and deduced from the division budget only after a while. Temporarily, only rough estimates are considered by accountants. In so far as the field is concerned, the final average cost is ultimately known at the division level, since the budget is issued division-wise. This implies that the sub-division officer (SDO) has no incentive to spare regarding the budget of the division; should he save, another SDO would enjoy the benefits.

The cost ultimately changed is unknown ex-ante, and does not ultimately result from the decision to replace, but rather results from the averaging of decisions taken in all divisions. Thus, decisions do not locally depend on techno-economic considerations.

Beyond these issues, the very nature of the budgetary system frames the decision-making. Field officers are not actually granted the so-called budgets they enjoy. On the contrary, for each decision, they have to file and get clearance for a technical sanction, and subsequently an administrative approval from the headquarters. Ultimately, they have to apply for a requisition of funds, which the headquarters has full authority to reject. In the operation of transformers, they have to remain within strict procedures of replacement on a one-to-one basis. Similarly, as far as development of new transforming capacity is concerned, they can only follow the centrally given instructions. They can in no way divert maintenance funds from capital expenditure, nor anticipate on investment, even if profitable in the mid run. Ultimately, since budgets cannot be diverted (not even managed by themselves), the field officers have to wait for transformers to burn to do something; and the replaced transformers are not always new ones. Indeed, while, trying to separate the budgetary responsibilities between divisional and sub-divisional levels, I inquired in my field surveys who pays for the transformer. The answer I received was 'nobody pays, provided nobody is responsible', implicitly referring to the circuit of administrative sanctions. Since the essential constraint is on rate of damage (on which field officers can easily argue that they have no means of action given the strict procedures they have to follow), once the damage has occurred, the cost of repair is of no importance.

Hence, decision-making at the local level is constrained by the internal organization of the boards. In the discussed example, the field would have the information on respective profitability of investment occurs places. Hence, under this centralized system, the headquarters is not able to take the right decisions, and the most profitable investments often remain undone.

More generally, SEBs are characterized by (*i*) the absence of budgetary delegation, which precludes taking right decisions at the field level, and (*ii*) a non-integrated reporting system. The term 'non-integrated' refers to the fact that the reporting system has hardly been designed to give precedence to costs over benefits, or to financial aspects over quantities and technical issues. Ultimately, SEBs rely on technical and administrative procedures that have been designed during the time of expansion and not under the present maturity phase. Therefore, new works and development of new capacities are favoured at the expense of more profitable decisions. The details of these decisions largely remain inaccessible to the headquarters because of the lack of an integrated information. But right to decide is neither allowed to the field officers since they lack a minimal managerial discretionary power. An organizational reform of SEBs is therefore a must for tapping the power sector's profitability.

## CONCLUSION

In India, some actions need to be undertaken so the power sector may turn profitable. PLF improvement and non-technical losses reduction may be worth with present tariffs, and could provide margins for an increase of 17 per cent of the energy level. These actions do not need unpopular measures such as non-technical losses reduction or tariff increase. This can be achieved either via the public sector, if able to turn from its former objectives and to manage a public reform, or via the private sector, if enough guarantees are given to it. All this without charging more the public budgets. The actual question then is to understand why these actions are not undertaken. One part of the answer might come from the general way SEBs are operated, based on self-enforcing political executive instructions, absence of focus on costs and budgets in actual decision-making, and absence of properly designed information system (Ruet, 2002 and 2005).

## NOTES

1. If no special mention, figures are drawn from the annual report on the functioning of SEBs, issued by the Planning Commission, 1997 and 2002.
2. A crore is an Indian unit, that represents 10 millions. Rs. 1 crore = Rs. $10^7$; \$ 1 = Rs. 42, at the value of 1997, and Rs. 45 at the value of 2004; so Rs. 1 crore = 2,40,000 dollers. The 'paise' is the cent of rupee. 100 paise = 1 rupee. We keep the value of 4 crore/MW constant over the period, which is true in current prices, and therefore in constant prices with a low inflation, and equally nearly true in dollars, with a parity that is at the same level as in 1997.
3. It is of 115.40 paise for SEBs, if one takes into account the share of energy produced by SEBs and the share of energy purchased from other utilities (Planning Commission, 2002). It was of 81.8 in 1996-7 (Planning Commission, 1997). It is 49.60 for NTPC (NTPC, 1997; data for 2001-2 is not available but this figure is stabilized). This gives a mean of 90.90 if one takes into account their respective weight in generation (Planning Commission, 2002; the figure was out of the same calculation of 71 paise in 1996-7).
4. Sources: New figures for Orissa: World Bank (1996), Gridco (1998); for Haryana, World Bank (1997), Quraishi (1998); for Andhra Pradesh, APSEB (1998); for unofficial figures of Haryana, Punjab and Delhi: interviews conducted from 1998 to 2001, matter confirmed officially by Planning Commission, 2002. Former figures: Planning Commission (1997).
5. Let us divide the network within the existing network and the additional network necessary to evacuate the additional energy. The average cost of supply includes an annual instalment for the use of the former. Whereas the annuity of the latter part would modify the average cost, its remuneration has already been computed under its investment form. The annuity of the existing network should be divided as well by a larger amount of energy. However, since this amount does not vary among different technical strategies, it doesn't influence the relative choice. In absolute terms, one factor is playing positively, the other negatively. Ultimately, the global sign of the variation on the average cost will depend on the evolution of the accounting capital ratio generation/network and the speed of growth (linked with amortisation). By and large, it can be assumed that with only 11 per cent of increase, both the ratio between new and old assets and between generation and transmission won't be affected much, and this problem is therefore of second order.
6. Considered as fixed from the time of reference for the calculation.
7. Since calculations work on an average basis, i.e. considering the whole installed capacity at 64 per cent. In reality, some plants, especially from NTPC are already above this level. Some electric regions, as west and south are respectively at a PLF of 70.2 and 75.8. This means that with

improvement of the others up to 68.5 per cent, the new average would even be more that this figure; the equivalent capacity margin is therefore higher than 4.4 GW. Improvement of plant utilization can also easily come from interconnection lines between electric regions in order to use the potential, at least, of the eastern region, that has a PLF of 42.2, arising not just from technical problems, but due to absence of interconnection. But the study of interconnections is beyond the scope of this paper, and we remain on the conservative side.

8. From linear extrapolations made after the results of repair and maintenance programmes presented in the annual report of the Ministry of Power, 1997. The result is straightforward. Proportionality is kept for 2002.
9. Several constraints might appear or be anticipated; an anticipation cost should therefore theoretically be considered, but would be of second order. Since we compute annuities, it is not taken into account.
10. All figures come from World Bank (1996, 1997 and 1999), Gridco (1998), annual reports of NPC from 1994-5 to 1997-8.
11. Haryana's sales being 3.2 per cent of Indian sales (Planning Commission, 1997).
12. Since Andhra accounts for 13.8 per cent of the national demand (Planning Commission, 1997).
13. Orissa represents 2.6 per cent of the demand (Planning Commission, 1997).
14. Only 128 GWh sold in 1997-8.
15. NPC invested Rs. 356 lakh in asset increase in 1995-6, for an increase of distribution capacity from 28 to 94 GWh, which gives Rs. 5.4 lakh per 'average MW' as a rough estimate of the development cost. In 1996-7 and 1997-8, assets increase were 242 and Rs. 216 lakh, and power purchase 107 and 128 GWh. Considering the growth of supply along with the development cost, one can consider that over this amount, Rs. 70 and Rs. 113 lakh were dedicated to supply growth. Hence remaining Rs. 172 plus Rs. 103 lakh for losses reduction.
16. Mathematically, let us model networks as being composed of N lines of same length and carrying the same amount of energy P. For total energy Q, the needed number of lines is given by N = Q/N. Hence, the rate of losses t is proportional to $Q/N^2$. Then, if Q is fixed (to measure the impact of losses reduction only), $\Delta\tau/\tau = -2\Delta N/N$, where $\Delta N$ is proportional to the investment. Hence, $\Delta N_{1\%, India} = \Delta N_{1\%, Noida} \times N/N' \times t/t' = 8{,}500 \times 8/22 =$ 3,100. Multiplied by 10, we are not far from the figure of Rs. 35,000 crore.
17. To be completely rigorous, integral calculation would be needed. We have carried this out. This leads to a 30 per cent increase in cost in the Noida case, and provides a 'lower limit' of a few percentage in the SEB cases. But first, the error margin in allocation of expenses might be from this order in Noida case, whose representativeness is not granted, and is therefore only used to provide an additional confirmation for the broad

estimation of the total cost. Second, in our allocation of investment programmes to technical losses reduction, we have made a majoration of the cost when purpose of investments was not clearly shown. Third, a 10 per cent improvement in losses has some scale-increasing returns, since minimal restructuring has already taken place; this is neglected here. All in all, the figure of Rs. 35,000 crore might be a upper limit of the actual figure. Hence, further results on the profitability of losses reduction are robust, and on the safe line.

18. This would of course not remain true at a particular time if one considers an increase in the usage intensity of the network, but this is true with a constant intensity of loads, in a network with homothetic growth.
19. Also, these three strategies do not lead in the long term to the same electric system (plants + networks) configuration. Then it could be argued that the value of use is different at the end of the period, and should be taken into account when comparing the various solutions. To that, however, we shall argue that the respective values of use can be considered as techno-economic equivalent, since they not only lead to the same availability of power but also present comparable value of use (or net present value). Indeed, enhancing the availability of the system revaluates the former assets, and makes comparable the various net present values, since these are based on future benefits. Then, neither value of use, nor any anticipation costs need to be introduced in the analysis. Comparison in the section on techno-economical discussion can rely on investment costs and annual cash flow only.
20. It is about Rs. 180 crore in Orissa for institutional development, technical assistance and training, according to World Bank (1996). This figured has been multiplied by the number of SEBs, since the consultancy cost of organizational change can be, in the first order, considered as constant over the SEBs.
21. Calculated by charging the 22 per cent level of non-technical losses at a rate of 173 paise per kWh.
22. As far as 2001-2 is concerned, we can consider that tariffs have already risen in the permitted range, progress has to come from managerial and technological improvement.

## REFERENCES

Government of India, 'Electricity Sector in India, Organization and Financing of Development', paper presented by M.R. Srinivasan in Paris, Planning Commission, January 1998a.

———, 'Managing Distribution Reforms', Ministry of Power, Committee on Distribution Working Document, 1998b.

———, Report of the N.K. Singh Committee on Financing the Power Sector in India, Planning Commission, New Delhi, 2004.

Gridco, 'Reform in Gridco—Revenue Improvement Action Program', Gridco Information Cell Document, Bhubaneswar, Orissa, 1998.

Ministry of Power, *Annual Report*, New Delhi, 2003.

Mohan, R., 'The India Infrastructure Report', special supplement of 1997 to the *JMJ Book Info*, New Delhi, 1997.

NPC, 1995, 1996, 1997, and 1998, Annual Reports of Noida Power Company Ltd.

NTPC, *Annual Report 1996-7*, New Delhi, 1997.

Planning Commission, *Annual Report on the Working of SEBs*, New Delhi, 1997.

———, *Annual Report for the Working of SEBs*, New Delhi, 2002.

Quraishi, S.Y., 'Power Sector Regulation and Restructuring: Haryana's Experience', paper presented at the International Forum on Power Sector and Restructuring, New Delhi, 1998.

Ruet, Joël, 'Winners and Losers of State Electricity Boards Reform', occasional paper, Centre de Sciences Humaines, New Delhi, December 2001.

———, 'A New Paradigm for the Indian Power Sector', in R.K. Mishra and G. de Bernis (eds.), *Privatization of Public Enterprises*, New Delhi: Manak Publishers, 2002.

Shukla, P.R., 'Developing Countries and Global Climate Change, Electric Power Options in India', report prepared for the Pew Center on Global Climate Change, Ahmedabad, Gujarat, 2000.

Staff Appraisal Report, Orissa Power Sector Restructuring Project, Report no. 14298-IN, 1996.

———, Project Appraisal Document, Haryana Power Sector Restructuring Project, Report no. 17234-IN, 1997.

———, 'Power Sector Reform and the Privatization of Distribution in India', internal working document prepared by H. Salgo et al., March 1998.

———, Staff Appraisal Document, Andhra Pradesh Sector Restructuring Project, 1999.

World Bank, 'Reforming the Power Sector: Controlling Electricity Theft and Improving Revenue, Policy', 4p, note by Bhavna Bhatia and Mohinder Gulati, Public Policy for the Private Sector series, note no. 272, World Bank, 2004. http://rru.worldbank.org/documents/publicpolicyjournal/272bhatia_gulati.pdf.

# 13

# *Electricity Reforms, Firm-level Responses and Environmental Implications*

DEEPA MENON-CHOUDHARY, P.R. SHUKLA,
TIRTHANKAR NAG AND DEBASHISH BISWAS

## ENERGY AND ENVIRONMENT LINKAGE

The production, transportation and consumption of energy resources, especially of fossil fuels such as coal, oil and natural gas, generate negative environmental externalities. Air pollution due to emission of greenhouse gases (GHGs) and local pollutants is a major externality from energy use and is the focus of this study. The GHGs, including carbon-dioxide ($CO_2$), methane ($CH_4$) and nitrous oxide ($N_2O$) act as a partial blanket that traps some of the thermal radiation from earth's surface and makes it substantially warmer than it would be otherwise. This causes global warming, which has various negative impacts on agriculture and food security, human health, sea levels, and coastal areas in the long run. The concentrations of local pollutants like sulphur-dioxide ($SO_2$), nitrogen-dioxide ($NO_2$), suspended particulate matter (SPM) and carbon-monoxide (CO) in the ambient air reflects the air quality in an area. When these concentrations exceed the ambient air quality standards, it results in direct and immediate damaging impacts on human health and sensitive ecosystems, besides having other local and regional impacts such as acid rains.

Economic growth and development calls for rapid growth of the energy sector of a country. This is because there is increasing demand for fossil fuel by major energy-intensive sectors, like power generation, steel, cement, chemical and fertilizer, transport, etc., and their use, in turn, contributes to high levels of GHG and local pollutant emissions. The challenge for policy-makers is, therefore, to achieve developmental objectives without adversely impacting the environment (Chandler et al. 2002). A synergy between economy and energy policies, energy and environmental policies, and local, national and global environmental policies is vital for national sustainable development as well as for effective environment management.

## Energy, Electricity and Emission Trends

In India, expanding industrialization, increasing incomes, rapidly rising transport and modernizing agriculture have led to rapid growth in energy use over the past decade. Indian energy use has grown faster than GDP for the last 20 years (Economic Surveys, 1990-2002; CMIE, *Energy*, 2003a). Most notable is the case of the electricity sector, where consumption has grown at a rate higher than GDP and energy for the past two decades and this trend has become remarkably pronounced in the 1990s (Fig. 13.1).

Energy and environment-related concerns of the Indian economy include the growing gap between the demand and supply of energy

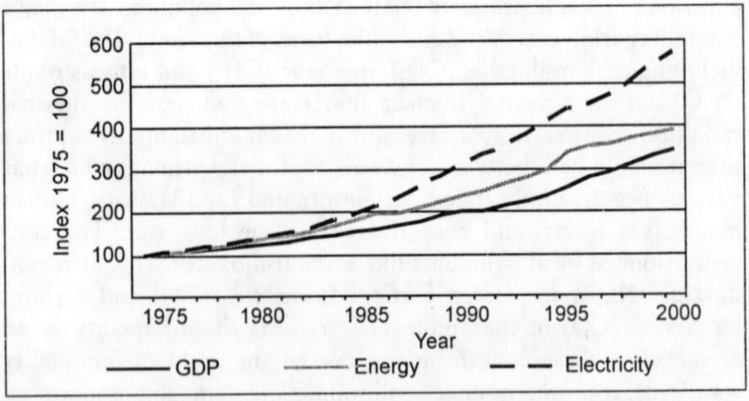


*Sources: Economic Survey of India* (1990-2002), Government of India; CMIE, *Energy* (2003a); CMIE, *National Income Statistics* (2004a).

FIGURE 13.1: GROWTH OF ENERGY, ELECTRICITY AND THE INDIAN ECONOMY

and the environmental externalities associated with energy use (GOI, 2000). Though the average efficiency of energy consumption is increasing, the energy intensity of the industrial sector remains high. The high fossil content of major energy fuels used in the end-use sectors contributes to high emissions, both at global and local levels. Thus, high growth of these sectors has resulted in a high elasticity of energy consumption and emissions with respect to GDP. The dynamics of transition from traditional to modern economy is expected to continue in the foreseeable future, further adding to the growth in energy demands (Shukla, 1997).

Coal has been the mainstay of the Indian energy system due to its vast indigenous reserves. And electric power generation is the dominant component of the Indian energy sector. The power sector consumes 40 per cent of primary energy, including nearly 70 per cent coal (CMIE, *Energy*, 1999). The generation mix is biased in favour of thermal, which is coal dominated. The share of coal-based power in gross generation from utilities has increased steadily from 54 per cent (in 1975) to 66 per cent (in 1985) and to 72 per cent (in 1995) (CMIE, *Energy*, 2001). The blue print of power sector development predicts that 52 per cent of the power generated in India in 2012 will still be from coal, while gas will be 15 per cent, hydro will be 27 per cent and rest will be renewables and nuclear (Ministry of Power, 2001).

The present growth rate of GHG and local pollutant emissions in India is above 5 per cent per annum (compounded) and the continued dependence on coal implies continuation of this trend (Garg et al., 2001b). This study focuses on $CO_2$ (GHG) and $SO_2$ (local pollutant), which are primarily emitted during coal consumption. In India, coal contributes to 75 per cent of total $CO_2$ emissions and around 63 per cent of $SO_2$ emissions (Garg et al., 2001b). Among the major emitters, large point sources (LPS), which include power, steel, cement and other industries, are the major contributors. In 2000, the LPS contributed to 65 per cent of all-India $CO_2$ emissions and 64 per cent of all-India $SO_2$ emissions. And, among the LPS, power plants are the major source, contributing around 44 per cent of all-India $CO_2$ emissions and nearly 45 per cent of all-India $SO_2$ emissions (Shukla et al., 2004c). Therefore, the power plants provide focused opportunities for emission mitigation in India.

## Policy Approaches for GHG and Local Pollutant Mitigation

Mitigation of GHG emissions for climate change control has important implications for national economy, but the mitigation of local pollutant emissions is a more urgent concern for national and local policy-makers. This is because the impacts of climate change are felt globally and over a longer duration, while local pollution causes immediate and direct impacts on health and ecology. Moreover, in developing nations like India, policies necessarily focus on more fundamental issues of socio-economic development such as poverty alleviation, creating basic conditions for human development and more recently on improving the local air quality. Policies and measures that specifically target $CO_2$ mitigation are more likely to happen in a global climate change regime.

In India, there are direct policies to control pollutant emissions from different sources, focussing on local pollutants. The ongoing policy reforms in different sectors, including electricity sector, which began in the early 1990s are also having implications on the air quality. But these reforms, including that in the electricity sector, have not always specifically addressed the environmental concerns. Policies and measures to manage air quality have not been integrated into the overall framework of electricity reforms; however, there have been indirect environmental implications. For instance, the changes in technology choices and efficiency improvements in power generation after the reforms have also resulted in the reduction in emission intensity of carbon and sulphur. This aspect of electricity reforms is brought out through case-studies of the states of Andhra Pradesh and Gujarat. These case studies provide an understanding of the varied firm-level implications of policy reforms already taking place in the electricity sector. But it is also necessary to align environmental concerns in the electricity reforms process, especially because the use of emission control technologies in power plants has implications on its fixed and operating costs and affects its competitiveness. Therefore, we also study the existing national-level policies for managing pollutant emissions from power plants and analyse alternate policies and measures that would control these emissions in a more cost-effective manner.

## STATE-LEVEL REFORMS AND FIRM-LEVEL RESPONSES

The power sector is the concurrent responsibility of both the central and state governments. The reforms, though initially implemented at the national level, have produced micro-level consequences at the state level which may be beneficial to both the local and global environment. Two states, namely Gujarat and Andhra Pradesh (AP), are chosen as samples to study the effects of reforms at a much disaggregated level of generating firms. The choice of the sample was guided by the progressive outlook of both states towards electricity sector reforms. AP started state-level reforms much earlier than Gujarat; and both states have a different kind of fuel mix. These factors would further benefit cross-case comparisons.

The study is based on primary data collected through questionnaire surveys and interviews with the utility managers and several subject-matter experts. The design consists of a comparative analysis of pre- and post-reform scenarios of utilities in the states, understanding the factors that drive the changes during reforms process and their effect on local and global emissions. Sulphur and carbon emissions have been used as examples of local and global emissions, respectively. The survey uses validated methods (Shukla, et al., 2004a, 2004b) and covers all the units of the utilities in both the states.

With the initiation of power sector reforms, a number of changes took place in the electricity sectors of Gujarat and AP. These changes occurred in various spheres: market structures, institutional and legal. The changes had significant effects at the firm level with respect to their fuel and technology decisions. At the firm level, the reforms have affected the fuel mix and technical efficiency, leading to changing emission intensity of generation. We explore the changing dynamics of these main drivers: fuel mix and efficiency at firm level and analyse their impact on the local as well as the global environment.

### CHANGING FUEL MIX

The fuel structure of generation in both the states has developed in a different manner due to the difference in resource endowment and geographical location of the states. Electricity generation in AP is dominated by thermal and hydro sources (AP TRANSCO 2001, AP GENCO 2003). The state has potential for hydropower generation due to the presence of a number of sources. Majority of the capacity owned by the state in 1990 was hydro; however, hydro generation has come down substantially since 2001. Coal technologies have shown higher growth mainly through the extension of existing coal plants. The state has easy access to coal from central and south-eastern part of the country. Post-reforms a number of private players have entered the generation industry. All the private generation is gas or naphtha based. Though most of the production of gas in India comes from the western offshore area, AP and Assam are the other producers of natural gas. In 2001, gas and naphtha-based plants accounted for 13 per cent of the total capacity compared to 1.3 per cent in 1990. The state has also been encouraging the development of wind power. There has been participation from the private sector with government support, yet capacities remain very low. As of 2001, the total wind capacity in the state remained at 89 MW.

On the other hand, power generation in Gujarat is mostly dependent on coal-fired stations. In 1990, three-fourths of its capacity was coal-based (CMIE, *Energy*, 2002a; CMIE, 2003b). Hydro stations are limited in number because of the low hydro potential of this region. Though the state is not endowed with enough coal reserves, it has access to coal from central and eastern parts of the country, which are rich in coal. Transportation of coal over long distances has made coal more expensive in Gujarat when compared to AP, which in turn has driven up its generation costs. The state has some lignite reserves, mostly concentrated in two locations (Kutch and Bharuch districts). Between 1990 and 2001, lignite production saw a steep increase of 109 per cent. The lignite reserves in the state have led to a number of lignite-based units being set up both by the state and jointly with other companies. The state also has reserves of natural gas. During 1990 to 2001, the production of natural gas in the state saw an increase of 87 per cent. In 2002, gas and naphtha-based plants accounted for 34 per cent of the total capacity compared to 3.5 per cent in 1990. During this period, nuclear capacity of 440 MW was added in the state. Though the share of coal has reduced, it remains the dominant fuel. The Centrally controlled companies like the National Thermal Power Corporation (NTPC) have also set up large gas-based capacities in the state. Gujarat, like AP, has encouraged the development of wind power, but the capacity remains very low.

In both states, a number of private players have set up capacities after the initiation of reforms. These private generators have used mostly naphtha or natural gas as the chosen fuel. Later, with steep hike in naphtha prices and increased production of natural gas in the states, private producers have been favouring natural gas over naphtha. As Fig. 13.2 shows, there has been a marked shift in ownership between 1990 and 2001, with the influx of private players. In both AP and Gujarat, the state capacities have grown at a slower pace. In Gujarat, the main entrants during the reforms period were private and Central sector gas-based plants along with the addition of a Centrally managed nuclear plant. AP also saw a huge influx of privately managed gas plants though their total capacities were much lower when compared to Gujarat.

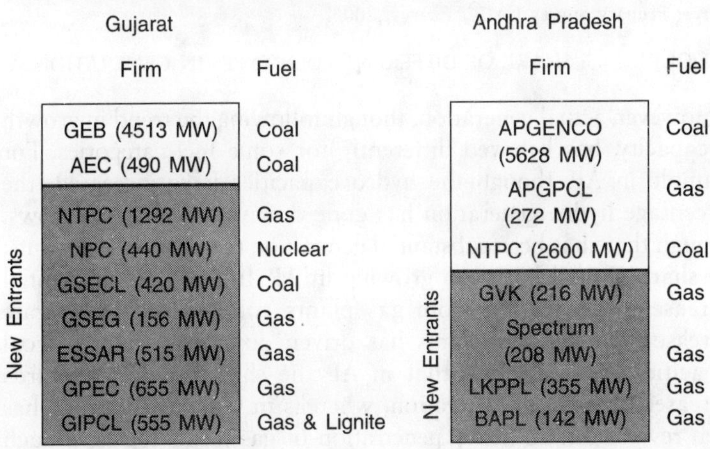


*Source*: Various utilities.

FIGURE 13.2: NEW ENTRANTS IN GUJARAT AND AP AFTER 1990

The proximity of Gujarat to natural gas sources and pipelines has led to much higher penetration of gas capacities. Compared to over 30 per cent of gas capacity in Gujarat in 2001, AP had only 13 per cent of its capacities based on gas. Even the centrally managed plants in Gujarat are based on gas compared to its coal-based counterpart in AP. Apart from this, low inflows in hydro reservoirs in AP have led to a very low utilization of its hydro capacity, with its contribution to total generation being only 12 per cent in 2001.

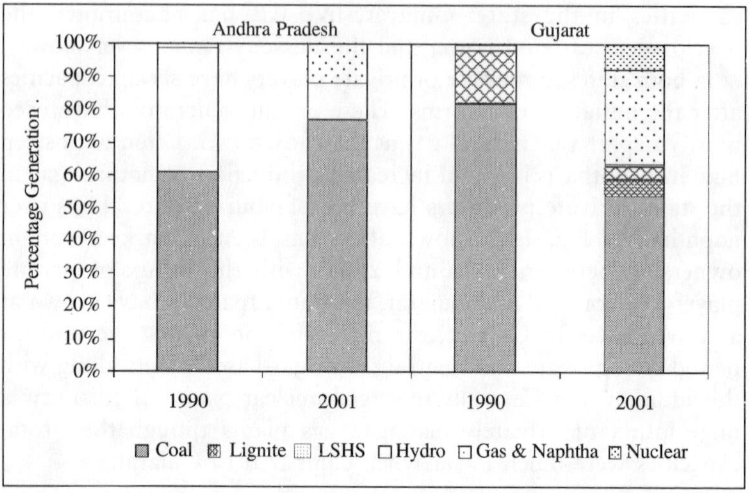


*Source*: Primary Survey CMIE, *Energy*, 2003.

FIGURE 13.3: SHARE OF DIFFERENT FUEL TYPES IN GENERATION

However, actual generation, though following the trend in growth of capacity, has behaved differently for some fuel categories. For example in AP, though the hydro capacities have increased, the percentage hydro generation has gone down due to poor inflows. Though there has been substantial generation from gas-based plants, the share of coal has been growing in AP. In Gujarat, substantial increase in generation from gas plants coupled with moderate increase from nuclear plants has driven down the share of coal capacities. Thus, we see that in AP the shift has been towards increased coal-based generation, whereas in Gujarat the trend has been reverse due to heavy penetration of gas technologies, though it has been offset to some extent due to the setting up of lignite plants in the state.

## TECHNICAL EFFICIENCIES OF THERMAL PLANTS

The entry of private players and changing fuel structure has been associated with improvements in energy efficiency of thermal plants. Heat rate,[1] a measure of thermal efficiency, has been used to study these changes. A number of factors influence the heat rate of a generating plant. In the planning stage, the choice of technology, size, and fuel puts a limit on the achievable thermal efficiency. Other factors like vintage of the plant also play a role in determining the upper limit of efficiency. This heat rate is specific to each of the generating units and is known as the design heat rate. The equipment manufacturer supplies the design heat rate. In practice, it has been found that the actual heat rate is higher than the design heat rate. Studies carried out on selected plants of Gujarat Electricity Board have reported significant deviations of the actual heat rate from the designed one (Alagh et al., 1998). A number of factors like ownership of the plant, management practices, maintenance schedule, and quality of fuel lead to this difference between actual and design heat rates.

A survey of current literature reveals that studies that examine policy incentives for adoption of energy efficient technologies (given that these efficient technologies help in reducing emissions) are relatively few in number (Verhoef and Nijkamp, 1999). Among these, a study in China (Zhang et al., 2001) focuses on the bifurcated development of efficient and inefficient technologies in China, factors that influence this development, and its impact on carbon emissions. Several studies focus on barriers to adoption of efficient generation technologies and its impact on global emissions in the Indian context (Khanna and Zilberman, 1999; 2001). Another study has reported a strong association of technical efficiency with unit size and capacity utilization (Singh, 1991). There are studies, as in Gujarat, that provide technical analysis of the causes of deviations between actual and design heat rates (Alagh et al., 1998).

The thermal efficiency of plants in AP varies from 1,900 kcal per kWh to 4,000 kcal per kWh. The heat rates of coal-based units vary from 2,290 kcal per kWh to 4,000 kcal per kWh. The thermal efficiency of plants in Gujarat varies from 1,900 kcal per kWh to 4,600 kcal per kWh. The heat rates of coal-based units vary from 2,300 kcal per kWh to 4,600 kcal per kWh. Indian coal has a relatively lower calorific value due to its high ash content of around 40 per cent, which influences efficiency to certain extent. In Gujarat, plants using lignite have heat rates over 2,500 kcal per kWh and those using low sulphur heavy stock (LSHS) have heat rates between 2,700 kcal per kWh to 3,400 kcal per kWh. In both states, recent technologies based on gas or naphtha have comparatively lower heat rates ranging from 1,900 kcal per kWh to 2,200 kcal per kWh.

In 1990, prior to reforms, plants having thermal efficiency in the range of 2,300-2,900 kcal per kWh supplied bulk of thermal generation in both states. In AP, plants having thermal efficiency in the range 2,300-2,600 kcal per kWh produced 31 per cent of the generation from coal and plants having thermal efficiency in the range 2,600-2,900 kcal per kWh produced 68 per cent. In 2001, plants having thermal efficiency in the range 2,300-2,600 kcal per kWh produced 31 per cent of the generation from coal. Gas generation had not penetrated in 1990 in AP; but after reforms, the total gas and naphtha capacity reached over 1 GW in 2001 and this entire capacity was added in the period between 1990 and 2001.

In Gujarat, coal plants having thermal efficiency between 2,300 kcal per kWh and 2,600 kcal per kWh produced 50 per cent of the generation from fossil fuels. In 2001, 70 per cent of the fossil fuel plants fell into this category. The average heat rates also have been declining for both gas and coal-based plants. Compared to coal-based plants, there were negligible gas/naphtha-based plants in the early 1990s. However, the total gas and naphtha capacity in 2003 reached over 3 GW and most of the capacity was added in the period between 1990 and 2003. Most of the gas or naphtha-based generation comes from units having heat rates between 1,900 and 2,100 kcal per kWh. There has been a shift towards lower heat rates. In 1998, 57 per cent of the generation was from units having heat rates between 1,900 and 2,000 kcal per kWh compared to 62 per cent in 2002. This has primarily been due to improvements in thermal efficiency of plants, particularly the central plants having higher heat rates. The narrow band of heat rates of gas and naphtha plants compared to plants using other fuels implies that technologies are almost homogeneous in terms of thermal efficiency and the variations that exist are due to other factors like size and maintenance.

With the shifting of coal-based generation to lower heat rates and penetration of gas and naphtha-based technologies, plants having comparatively lower heat rates are producing a major proportion of electricity. This implies that this pattern of growth is beneficial for the global climate, although many of the units, especially lignite and LSHS-based ones in Gujarat and the older coal plants in both the states are having high heat rates (compared to that in developed countries).

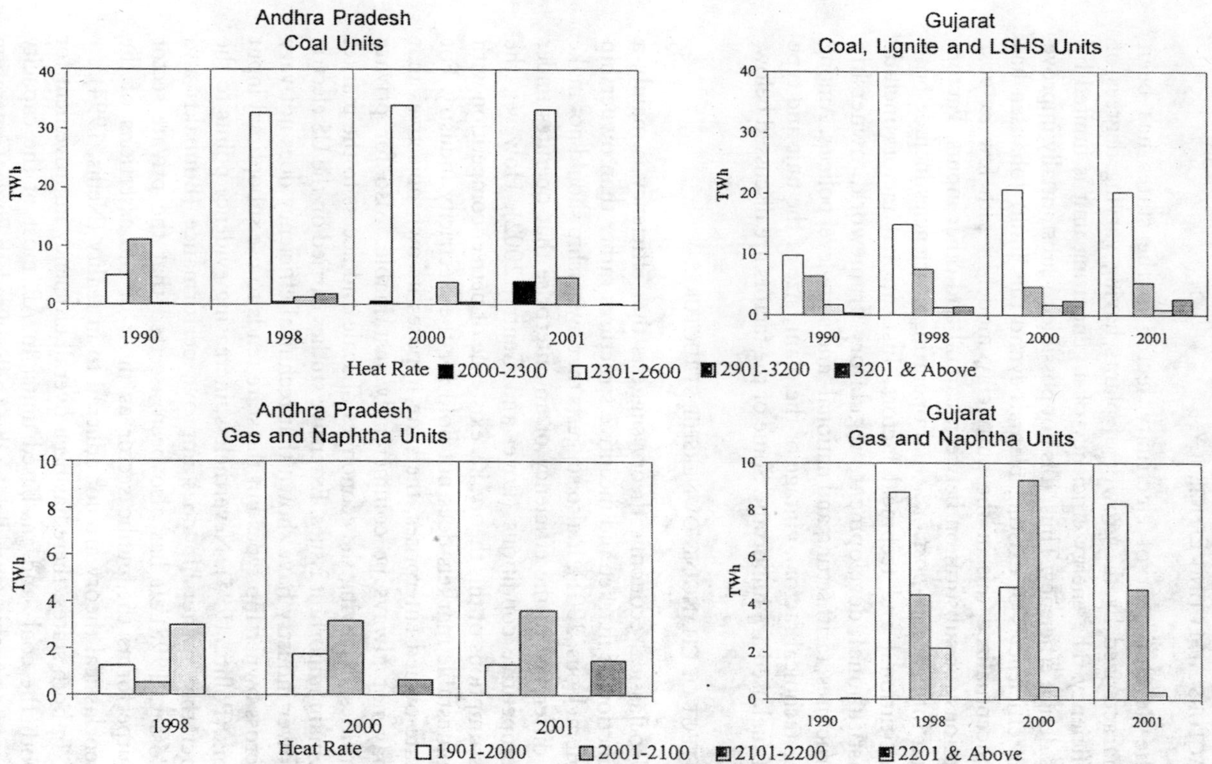


FIGURE 13.4: CLASSIFICATION OF GENERATION UNITS

## Factors Affecting Efficiency

The development of electricity sector in the states has been characterized by the growth of plants having varying degrees of efficiency. The energy efficiency of a generating unit is limited by its design efficiency. Thus, it is not possible to substantially improve the efficiencies of existing plants beyond this limit. Johansen has conceptualized this in the putty-clay model,[2] with few *ex-post* options for substitution between capital and other inputs. Factors that influence *ex-post* or actual operating efficiencies include both institutional and technical factors. Institutional factors include institutional development, regulatory framework, ownership structures, tariff setting and national- and state-level policies. Among the technical factors, vintage of the plant, size of the unit and type of fuel used are important. Some of the factors are discussed below.

### EFFECT OF CHANGING OWNERSHIP STRUCTURES

Neoclassical economic theory suggests that efficiency is mainly a function of market and incentive structures rather than ownership (Heracleous, 2001). It also suggests that ownership should not make a difference to the efficiency of an enterprise that operates under competitive conditions (Lovei and Gentry, 2002). However, the assumptions of this theory, i.e. existence of perfect competition with contestable markets does not hold for the electricity industry. For state-owned enterprises, the state imposes its own objectives that may not always be commercially sustainable. Also for private utilities, achieving the economies of scale and access to the market remain important issues. Extensive studies carried on the US electric power industry have shown that technical efficiency does not vary across ownership types but the state tends to use suboptimal input mix (Pollitt, 1995). Nevertheless, increasing evidence points to the fact that ownership does improve performance (Boubakri and Cosset, 1998). It has also been asserted that the private sector outperforms the public sector as most of the assertions of the neoclassical theory are not adhered to in reality (Nellis, 1994).

In both states, there has been overall efficiency improvement for both the coal- and gas-based units. In AP, most of the capacity based on coal is controlled by the Andhra Pradesh Power Generation Corporation Limited (AP GENCO), which is under state control.

In 2001, the most efficient (2,000-2,300 kcal/kWh) units, accounting for 9 per cent of the capacity, were under state control. Two recently installed units by the state are responsible for this; therefore, the high efficiency may be attributed to recent vintage. The bulk of electricity was supplied by the next range of efficient units (2,301-2,600 kcal/kWh) accounting for over 77 per cent of the capacity. This category included both Central and state owned units, with Central units having considerably higher efficiencies than the state units. This has been possible due to sustained renovation and modernization. In case of gas, most of the plants having higher efficiencies are controlled by the private sector. The existing gas contracts in the region are 'take or pay' contracts with one-sided liabilities for power generators. This coupled with the problems with supply of gas have aggravated the situation for gas units.

In Gujarat, most of the capacity based on coal, lignite or LSHS are controlled by Gujarat Electricity Board (GEB), which is under state control. In the year 2001, some of state owned shifted to a higher efficiency range, accounting for more than 60 per cent of the installed capacity in this segment. This is again due to sustained renovation and modernization exercises carried out by the state. The central and private sector control most of the gas plants having higher efficiencies. The same shift has been noticed for gas/naphtha-based units of the centrally managed plants, while some of the private gas plants have fallen back in terms of efficiency. Better management of fuel has been cited to be the single most important instrument in achieving higher efficiencies.

## UNIT[3] SIZE SELECTION AND ITS EFFECT ON EFFICIENCY

The electricity sector in both states is characterized by a wide variation in unit sizes. The unit sizes vary from 30 to 500 MW for coal-based plants in AP and from 60 to 210 MW for those in Gujarat. For gas-based units, the range of unit size is from 100 to 655 MW. The heat rate increases with reduction in average unit size (Shukla et al., 2004a; 2004b). The effect is more prominent for coal generating plants. However, smaller units of capacity less than 100 MW influence this effect. Beyond this size, the unit size does not have any significant impact on the heat rate. For gas-based units, the relation between unit size and efficiency is not that apparent and the heat rates lie within a narrow band irrespective of the unit sizes.

The above observation raises questions about the reasons for choosing such widely varying unit sizes. Interviews with plant mangers suggest that in case of coal-based units lower sizes were deliberately chosen for three reasons. Firstly, instead of a large unit, a set of smaller units were often chosen for better reliability. In the face of supply constraints and deficits, it was always considered prudent to invest a portion of the capital in smaller units even though they have less efficiency and have higher per MW installation and maintenance cost. Secondly, to avoid partial loading and achieve better efficiency, some of the stations were conceived as multiple sets of smaller units. This gave the advantage of shutting down one unit and operating others at full load for better efficiency. Thirdly, the flexibility of shutting down a single unit also offered scope for planned maintenance without much disruption of supply.

## TECHNOLOGY CHOICE AND PLANT VINTAGE

The oldest plants in operation in AP were installed in 1965. These units had lower unit sizes of 30 MW. The coal plants in AP are characterized by unit sizes between 200 and 250 MW after 1980. The oldest plants in Gujarat were installed in 1964. These units were based on LSHS and had unit sizes of 63 MW. In Gujarat, the growth of coal-based plants is characterized by two distinct range of unit sizes of 200/210 MW and 110/120 MW. Plants installed in the early 1960s and 1970s in both states had components imported from the USA. The units installed later had components mostly from Bharat Heavy Electricals Limited (BHEL) due to policies of the central government that encouraged setting up indigenous units. In addition, the unit sizes were constrained by the choice available with the manufacturer. With the advent of central sector plants, the unit size has further increased to 500 MW in AP. Central sector plants in Gujarat are gas based due to the central policy of installing plants only near the coal mines.

Naphtha- and gas-based plants have been set up in the early 1990s, with unit sizes around 50 MW in AP and 100 MW in Gujarat. Later, the unit sizes increased as improvements in technology made smaller sizes economically unviable. However, in recent years, the size of gas plants has decreased due to the entry of smaller players having investment constraints and different risk profiles. In addition, economies of scale are not that prominent for gas turbines. Considerations of reliability and loading are taken into account as most of the gas-based plants have multiple gas units with an associated steam unit for combined cycle operation.

In the case of coal-based generation, state-owned plants show higher thermal efficiency compared to the private sector. However, it cannot be concluded that state ownership leads to better overall efficiency in case of coal-based generation. Moreover, higher thermal efficiency does not always translate into lower cost of generation as is seen in the case of some of the private and Central sector units, which have lower cost of generation compared to state plants. The reforms also attracted private players who had favoured gas-based generation due to its low capital costs, reduced gestation periods and scale economies being achievable at much smaller sizes. But in recent years gas-based plants have been facing problems related to the availability of gas. Compounded with the take or pay contracts offered by gas suppliers, some of the plants have shown lower capacity utilization and reduced efficiencies. Ensuring gas supplies and enforcing contracts remain among the key issues that have to be resolved to further the growth of gas-based capacity as envisaged by the states.

## Impact on Emissions

### LOCAL POLLUTANTS ($SO_2$)

The development of electricity sector in the two states is characterized by a shift in the fuel structure with a growth of gas-based generation. This shift in fuel structure was accompanied by a change in efficiency of generation for both the gas- and coal-based plants. These changes were prompted by the policy of the central government allowing private sector participation in electricity generation and setting up independent regulatory mechanisms for this sector. These changes in generation have an impact on both the local and global emissions and the baseline determination as discussed subsequently.

The data collected provides generation, fuel consumption and fuel characteristics for all generating units in the two states for four years from 1990 to 2001 (1990, 1998, 2000, and 2001). Using the data, $SO_2$ intensities for the electricity generation industry are computed for these years. The overall $SO_2$ intensity computation takes into account the fact that emissions from the existing stock are heavily weighted by older inefficient plants. The marginal baseline is constructed using only the additional capacities for each year after 1990. Considerable difference between the overall and marginal baselines reflects the changing technology and fuel choices and raises the question regarding the choice of baseline for emissions trading.

AP, unlike Gujarat, is primarily coal based with a number of gas-fired unit coming up after the reforms. Fig. 13.5 shows that the overall $SO_2$ intensity for AP has increased from 3.5 g/kWh to 4.14 g/kWh from 1990 to 2001. A major contribution to this increase has been the decreasing generation from hydro plants. The contributions from hydro plants have decreased from 39 per cent of the total generation in 1990 to only 12 per cent in 2001. The $SO_2$ intensity for coal plants has decreased from 5.6 g/kWh to 4.9 g/kWh. This can be attributed to efficiency improvements through renovation and modernization and also to the addition of recent technologies having better efficiency. However, the marginal intensity curve has shown a phenomenal decline to 2.5 g/kWh mainly due to the influx of gas technologies having negligible $SO_2$ emissions and also due to improved heat rates of new coal installations.

In Gujarat, a reduction of 43 per cent in the overall intensity has been achieved between 1990 and 2001. The overall $SO_2$ intensity stands at 3.5 g/kWh in 2001. A major part of this decline can be attributed to the changing fuel mix of the state. The marginal intensity has declined to 1.55 g/kWh, much lower than AP, again

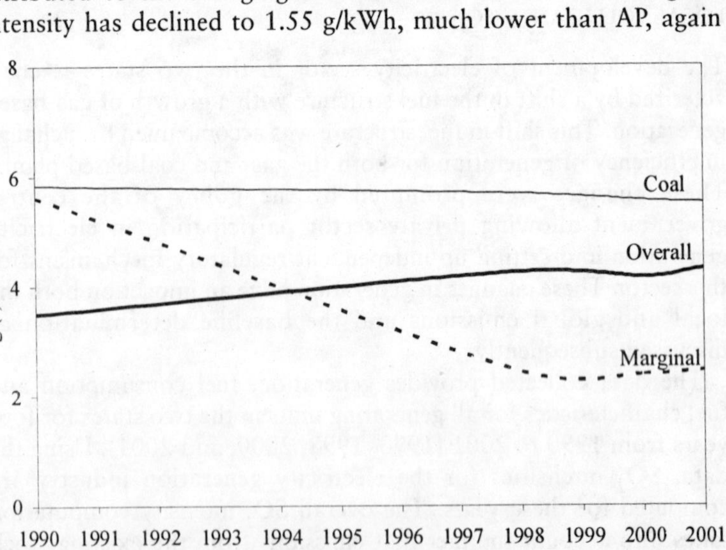


FIGURE 13.5: $SO_2$ BASELINE FOR ELECTRICITY INDUSTRY IN ANDHRA PRADESH

due to the heavy penetration of gas technologies compared to AP. Coal-fired units have shown a moderate decline of 7 per cent over the same period. Lignite plants have more or less maintained a steady intensity level. However, there was a dip in 2001, due to the installation of advanced Atmospheric Fluidized Bed Combustion (AFBC) based lignite technologies by a cooperative generation plant. LSHS plants, which are the oldest plants in existence in Gujarat, have lower heat rates and higher emissions. These plants have shown further deterioration over the reforms period, and the $SO_2$ intensity has gone up by 3 per cent and now stands at 12.2 g/kWh. In Gujarat, plants have problems with availability of natural gas. Many of the plants are dual-fired and use expensive naphtha when natural gas is not available. However, emissions from these plants are very low.

The above $SO_2$ intensity (Fig. 13.6) baselines show that there are considerable differences across regions with entirely opposite behaviour of overall baselines in the two states. In AP, the overall baseline is increasing while in Gujarat the overall baseline is

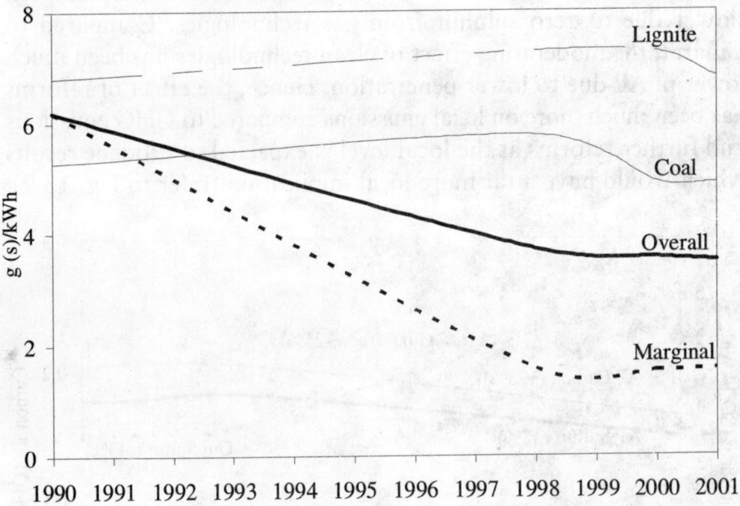


FIGURE 13.6: $SO_2$ BASELINE FOR ELECTRICITY INDUSTRY IN GUJARAT

decreasing sharply. Marginal intensities show a much sharper decline in both states. In regions using a varied fuel mix like Gujarat, there are considerable differences across fuels and also across technologies. All these factors highlight the problems of fixing an effective baseline for $SO_2$ trading.

## Comparison of $SO_2$ Emission Baselines and $CO_2$ Emission Baselines

Both $SO_2$ and $CO_2$ emission baselines depend on fuel consumption and are expected to follow similar trends. However, differences arise due to the changing mix towards cleaner fuels and use of efficient emission capture devices. In Gujarat, there have been significant differences in percentage decline of the overall baselines for carbon and sulphur. Carbon baselines have declined mainly due to the induction of gas technologies having lower carbon emissions compared to existing coal technologies. Gas technologies emit low carbon and no sulphur. Therefore, the difference between coal and gas technologies is still starker when it comes to accounting for $SO_2$ and the $SO_2$ baseline for Gujarat has decreased much sharply compared to carbon baseline. The same trend is seen in AP.

AP had an increasing overall baseline to start with due to steady decline in emission-free hydro generation. The sulphur baseline follows the same increasing trend; however, the increase has been slower due to zero sulphur from gas technologies. Compared to Gujarat, the moderating effect of clean technologies has been much lower in AP due to lower penetration. Hence, the effect of reforms has been much more on local emissions compared to GHG emissions and further reforms at the local level is expected to produce results which would have a far more local implication (refer to Fig. 13.7).

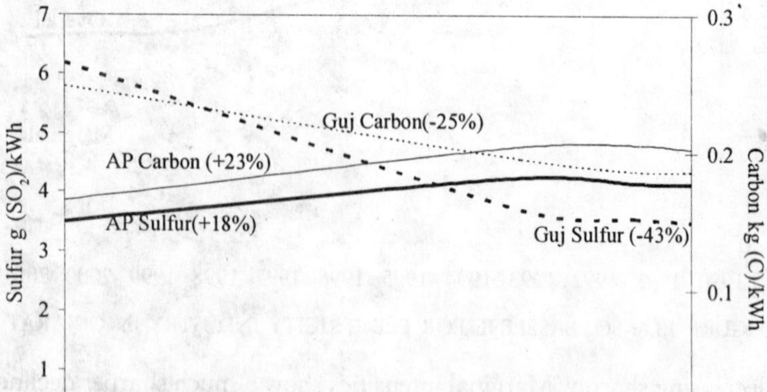


FIGURE 13.7: OVERALL EMISSION BASELINE FOR AP AND GUJARAT

## ALIGNING ELECTRICITY REFORMS AND ENVIRONMENTAL POLICIES

As seen in the previous section, electricity reforms have had indirect environmental implications in terms of decline in marginal emission baselines, more so in the case of local emission baseline. Direct environmental policies are also being adopted for controlling air pollution in the country, focussing primarily on local pollutants. These policies rely on command-and-control instruments such as emission norms and specified technologies to achieve their objective; while market-based instruments such as emission tax, emissions trading and others have been rarely applied. This is true for the electricity sector too, where emission norms and specific technologies are being mandated to control the local pollutant emissions. The choice of instruments, whether command-and-control or market-based, influence the characteristics of emission control technologies/processes adopted by the power plants, which in turn has implications on their costs and competitiveness. Given these strong environmental linkages of the electricity sector, it is necessary to adopt suitable policies and instruments for managing the air quality and align them with the electricity reforms process.

### LINKAGES BETWEEN $SO_2$ AND $CO_2$ CONTROL

Studies (Garg et al., 2003) have shown that while GHG and local pollutant emissions in India are connected presently, however, in future there is a disjoint between the two and they do not move in synchronization. The use of low-sulphur fuels in the transport sector, rapid penetration of sulphur removal technologies in the power sector and large industry boilers will enable early decoupling of the two emissions in the reference case (Pandey and Shukla, 2003). The co-benefits of a sulphur mitigation policy and carbon mitigation policy are likely to be asymmetric, since carbon emissions are lower in a carbon mitigation policy than in a sulphur mitigation policy; but a carbon mitigation policy would have greater impact on sulphur emissions reduction. This is because energy substitution (natural gas and renewables substituting coal) in a carbon mitigation policy would also reduce sulphur emissions. Thus, there would be cost sharing in a carbon mitigation policy. But a sulphur mitigation policy would introduce technologies that abate $SO_2$ emissions, with little impact on $CO_2$ emissions.

Presently, India does not have any commitment for GHG emissions mitigation under the Kyoto Protocol. And, unless there is a global carbon mitigation market, it would be uneconomical for India to invest in technologies that have greater carbon mitigation since it would adversely affect the use of its primary resource, namely coal. The financial constraints of the country make it imperative to use indigenous resources optimally and in priority areas. Therefore, till a carbon market develops, a $SO_2$ mitigation policy should be followed to promote the use of coal by adopting clean coal technologies, improved coal quality and other emission control technologies such as Flue Gas Desulfurization (FGD) that would reduce sulphur emissions. Alternate technologies like gas and renewables would in any case penetrate in a global climate change regime when a market develops for carbon emissions.

## $SO_2$ Control Policies in India

The $SO_2$ emissions from power plants are presently controlled using measures specified by regulations. These include policies that specify the stack height criteria for dispersion of $SO_2$ emissions. After 2000, there have been other policy directives regarding the implementation of emission control technologies. For instance, new coal power plants above 500 MW (and not using Integrated Gasification Combined Cycle (IGCC)[4] generation technology are required to keep space for installing flue gas desulphurization equipment (CPCB, 2003). This directive implies that the use of FGD technology would be made mandatory in the future. Another policy decision has made it mandatory to wash coal used 1,000 km away from the mine mouth or in critically polluted or urban areas (CPCB, 2003). This measure, aimed at reducing fly ash and improving efficiency, would also reduce some sulphur.

However, these measures rely on technological solutions that are brought through regulations and all plants have to comply with them. Such technology push policies do not consider the differences in marginal abatement costs of plants, thereby limiting the flexibility of plants in abatement choices. They are based on the relative ease of implementation rather than on any economic criteria. However, there are instruments such as emissions trading that would complement the regulations and control emissions in a cost-effective manner in the long run. In this context, we analyse the implications of an

$SO_2$ emissions trading system *vis-à-vis* a technology push policy for controlling sulphur emissions in the long run.

## Opportunities for $SO_2$ Trading in Indian Power Plants

Emissions trading is defined as follows: 'Essentially, a properly designed emission trading program is a form of environmental regulation that allows a group of sources to reach a specified emissions target at lower cost' (UNEP, UCCEE and UNCTAD, 2002). There are three forms: 'cap and trade', 'baseline and credit' and 'offsets'. In the 'baseline and credit' programme, the participants define their baseline and if actual emissions are below the baseline they earn credit that can be traded. The 'offset' programmes are used to compensate for the additional emissions from a new source or expansion of an existing one. In this study, we focus on the cap-and-trade system (also known as allowance trading). This approach aims to control emissions from a group of sources at costs lower than if the sources were regulated individually. Rather than specifying individual control requirements, this approach sets an overall cap that is the maximum amount of emissions per compliance period. Authorizations to emit are then allocated to participating sources in the form of emission allowances through some mechanism (freely in proportion to historical emissions or performance standards and/or through auctions), with the total allowances not exceeding the cap. The sources must accurately measure the emissions and turn in the same number of allowances as emissions at the end of the compliance period.

A well-designed cap-and-trade programme is a form of environmental regulation that not only controls emissions, but also creates significant cost savings. The cost savings is achieved in two ways. Firstly, the regulator does not impose specific emission control technology on each source while specifying the target; therefore, sources have the flexibility to determine their own emission control strategy. Secondly, if sources are able to reduce emissions at relatively low marginal costs, they could implement larger reductions, since they gain by trading with those firms that incur higher marginal abatement costs. And firms facing high marginal abatement cost can lower their own compliance costs by purchasing the allowance from firms with lower costs. Thus, there would be cost savings compared to a purely regulation-based approach. The flexibility in choosing strategies also creates incentives to go in for higher R&D to develop lower cost technologies (UNEP, UCCEE and UNCTAD, 2002).

The development of an emissions trading system depends on the presence of a significant number of emitters having varying control costs. Past research (Kapshe, 2003) has shown that 82 large coal-based power plants in India contribute to nearly 50 per cent of all-India $SO_2$ emissions. These plants are characterized by differences in terms of performance, quality of coal used, generation technologies, efficiency of operations and maintenance, emission control measures, efficiency factors, plant ownership and region of operation, the age of the unit and so on. These differences lead to differences in the marginal abatement costs between plants. For instance, the average age of the power plants in India is 13 years and these older power plants emit more $SO_2$ per unit of power generated (Kapshe, 2003). The marginal abatement costs would be higher for older units as compared to newer units that are more efficient. Again, many plants near coastal areas use imported coal. Imported coal, while low in ash content, has higher sulphur content than Indian coal, which implies the need to invest in emission control measures. These factors create differences in emissions and abatement costs among plants and provide a good opportunity for developing an emissions trading regime.

### SIMULATION

A bottom-up energy-environment model, the AIM/Local (Asia-Pacific Integrated/Local Model) is used for making future projections (till 2030) of $SO_2$ emissions in different scenarios. It is also used to analyse counter-measures and their impact on future emissions. This model follows a linear programming approach to find an optimal solution by selecting a combination of technologies with the least cost while satisfying the given constraints of fulfilling the demand and meeting environmental targets (Shukla et al., 2004c). Emission calculation methodology used in AIM/Local Model is in line with the recommended methodology of the Intergovernmental Panel on Climate Change (IPCC, 1996).

The data for power plants in India are obtained from diverse data sources, including published documents of Government of India, state governments, government organizations, industry federations and autonomous organizations (Garg et al., 2001a, 2001b, 2002a, 2002b; Kapshe, 2003). The future LPS data is taken mainly from published reports and databases like CMIE (2002a, 2002b). These provide status information of the planned investment projects in India for the power sector till almost 2010. The projections for power plants for the next 30 years have been made based on retrofitting and capacity augmentation options for existing plants, present policy directives of the government and expert opinion.

Scenarios are developed to analyse alternate policy instruments for controlling $SO_2$ emissions from coal-based power plants in India. These scenarios are developed for the business-as-usual case (BAU), which is a dynamics-as-usual and assumes continuation of macro-economic trends (including structural changes in the economy), demographic and energy sector trends (like autonomous energy efficiency improvements and penetration of clean technologies and renewable fuels), and policy trends, including those for power plants (see Fig. 13.8).

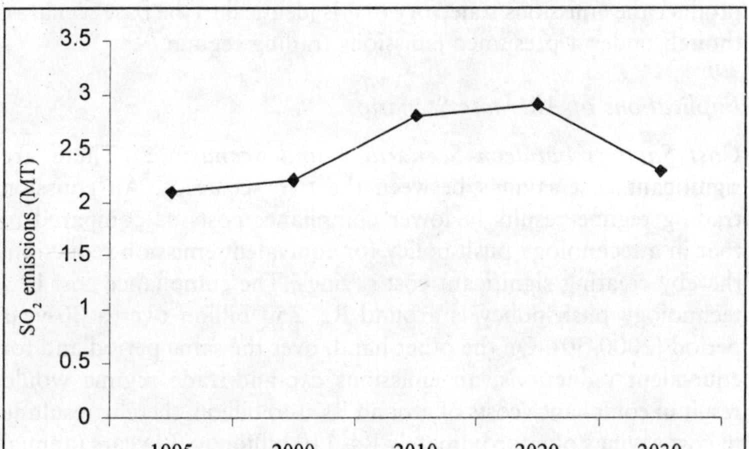


FIGURE 13.8: $SO_2$ EMISSIONS FROM POWER PLANTS IN THE BAU SCENARIO

Policy interventions that drive the trajectory in BAU could be brought about either through regulations that specify the technology or through an emissions cap-and-trade programme that leaves the technology choice to the plants. In a perfect market, the costs and path of adopting pollution control technologies should be optimal in both these scenarios. However, due to imperfections in the system this does not happen in actual situations. The present study develops scenarios for the two alternate policy pathways, namely

*Scenario 1: Technology Policy Scenario*

This scenario is modelled assuming that policy directives introduce technology policy instruments such as recommended technology types and standards. Under this scenario, the power plants do not have the option of choosing between alternate technologies or no abatement to minimize their abatement costs.

*Scenario 2: Technology Policy Equivalent Emissions Cap-and-Trade Scenario*

This scenario assumes that an emissions cap-and-trade instrument is adopted to reduce $SO_2$ emissions. The model is calibrated to produce the emissions trajectory that is identical to the base scenario, though under a presumed emissions trading regime.

*Implications of Alternate Scenarios*

*Cost Savings between Scenario 1 and Scenario 2:* There are significant cost savings between the two scenarios. An emission trading regime results in lower compliance costs as compared to that in a technology push policy for equivalent emission reduction, thereby creating significant cost savings. The compliance cost in a technology push policy is around Rs. 250 billion over a 30-year period (2000-30). On the other hand, over the same period and for equivalent reductions, an emissions cap-and-trade regime would result in compliance costs of around Rs. 140 billion, thereby resulting in cost savings of approximately Rs. 110 billion in 30 years (annual average of Rs. 3.6 billion).

The cost savings arise due to the inherent characteristics of an emissions trading instrument. For instance, regulations that specify control measures do not distinguish between new and old plants. New regulations place few costs or burden on plants that choose to operate older and dirtier units, but require those building new units to install expensive controls. This is seen in the policy directive that requires new plants with capacities above 500 MW to keep provision for FGD installation. This creates several distortions. For instance, advantages of scale economies would be lost if firms decide to set up plants with capacities of less than 500 MW. It also encourages continuing operation of older and dirtier units. But this is not possible in an emissions trading system, where plants can emit only according to the allowances they hold. In case they are unable to do so, they would have to purchase allowances. Another reason for the higher compliance costs in a technology push policy is the limited flexibility offered by this policy. A technology push policy would require installing the best among the different grades of de-sulphurization technologies. This limits the flexibility of plants to choose between the alternate grades depending on their cost structure.

At a broader level, the differences in compliance costs between the two scenarios of the base case arises because regulations are developed more on a policy basis, where the focus is on broad political and social considerations; whereas business decisions are made on economic considerations, where small differences in cost and regulatory design make an enormous difference. An emission trading system tries to achieve a balance by aligning economic and environmental incentives, thereby making it less expensive for the plants to comply.

*Allowance Price in Scenario 2:* In an emissions cap-and-trade regime, the allowance price in a year reflects the marginal abatement costs of $SO_2$ reduction in that year. The marginal abatement cost is around Rs. 6,825 per ton of $SO_2$ in 2005 and is expected to stabilize at around Rs. 5,000 per ton of $SO_2$ in 2030. It is high in the initial years, when the cap is applied, due to the investments in technologies that are required to meet the allocated emissions as well as other transition costs. However, the increase in demand for alternate technologies and increasing competition would lower the cost of technologies, thereby reducing the marginal abatement costs for plants in later periods. The replacement and retrofitting of existing sulphur intensive technologies by cleaner technologies in initial periods generates a momentum that reduces costs in the later period.

## Designing an Emissions Trading System

### Programme Structure

Power plants in India offer substantial potential for sulphur mitigation since they contribute to nearly half of all-India $SO_2$ emissions. The differences in abatement costs among plants would promote emissions trading. Since the number of plants is limited, locations concentrated and they are mostly in public domain, it should make enforcement and monitoring easier.

#### *Setting Emission Caps*

Emissions cap has to be set based on detailed cost-benefit analysis. This analysis would be based on epidemiological studies that estimate the health impacts of pollutant concentrations in the ambient air and studies related to other impacts such as that on property and ecology. This coupled with cost estimates would enable policy-makers to decide the required levels of emission reduction and set the aggregate emissions cap. The present paper, however, does a cost-effectiveness analysis between alternate policy scenarios for the BAU scenario; therefore, the proposed emission cap is set to follow the BAU trajectory for $SO_2$ emissions.

#### *Allocation of Allowances*

The allocation of allowances could be based on historic emissions, on performance parameters such as heat rate, sold through auctions or a combination of these methods. A mix of these methods would be more effective than a single mechanism for several reasons. Firstly, it will prevent the system from penalizing more efficient plants; secondly, it would prevent large emitters from getting benefits; thirdly, it would give due consideration to new plants that would come in the future; and fourthly, revenues would be obtained from auctions.

#### *Monitoring of Emissions*

Accurate and continuous monitoring of emissions from individual sources is very important in a cap-and-trade programme and a repository for the emissions data from the different units has to be maintained. Existing measurement methods have to be improved and the measuring equipments in power plants have to be standardized. Certification agencies would be required to certify the equipments as well as emissions reported from the units. The credibility of these agencies is crucial for the success of an allowance trading programme. These agencies could be under private parties duly certified by the government to instill confidence based on their independence and technical competence.

### *Accounting of Transactions*

An accounting system is required to track the holdings of allowances, to record transactions and at the end of the year give the details to authorities who would then compare the annual emissions with allowances held by the plant. Strict penalties have to be imposed on defaulters so that they act as a deterrent against false reporting.

### *Utilizing Savings Generated*

The savings generated in the emissions trading system should be used for supporting this system by investing in the infrastructure required for monitoring and enforcement and strengthening implementing institutions. It should also be invested in clean coal technologies and for improving the air quality in hotspots. Thus, there is a good opportunity for using economic forces to control $SO_2$ emissions from power plants optimally and using the savings generated for additional air quality gains, making it a socially cost-effective solution.

### *Implementing Stringent Local Air Quality Standards*

An emission trading system has to be implemented in coordination with local air quality regulations. This is because this system imposes only a national cap and no control on the local intensity of pollution, which would allow polluters to acquire an inordinate number of permits from the market to cover their excess emissions, thereby creating hotspots. Moreover, the presence of existing infrastructure and raw material creates path dependencies, resulting in setting up of new plants in the vicinity of older plants or expanding existing plants, which would aggravate the problem of existing hotspots. This path dependency has to be reduced to prevent deterioration of the air quality, especially in hotspots. These problems can be prevented by adopting local air quality standards that are more stringent than the national ambient air quality standards. This would make it impossible to purchase extra allowances to cover excess emissions and promote the spread of new plants to other locations. The State Pollution Control Boards (SPCB) in India have the legal authority to set more stringent emission norms than the national standards in hotspots.

### SPILLOVER BENEFITS

An emissions trading regime would have significant spillover benefits. For instance, a cap-and-trade system creates incentives for far greater reductions in emissions, thereby spurring technological innovation and energy efficiency. This would lead to investments in R&D to invent and commercialize new technologies, promoting technological breakthroughs. This regime could also provide impetus to the development of an environmental industry in India to supply pollution control technologies. For instance, currently India has 21 large coal washeries but very few are devoted to power requirements due to the huge investments required. An emissions trading regime would create competitive markets and make it economical for setting up more washeries since it would increase demand for clean coal.

### CHALLENGES

#### *Institutional Issues*

The critical issue in developing an emissions trading system in India is the question regarding which authority would be responsible for managing this system. Two ministries that are directly involved would be the Ministry of Environment and Forests (MoE&F) and Ministry of Power, though there are other ministries such as the Ministry of Coal, Ministry of Petroleum and Natural Gas, which would also be affected by this system. The organizations for implementing the environmental policies come under the MoE&F and include the Central Pollution Control Board (CPCB) and the different SPCBs. Given the fact that the environmental management is multi-sectoral and given the presence of several ministries in the energy sector and other ministries and organizations for environmental management, there is a need for a coordinating mechanism to manage the process.

Strong monitoring and enforcement institutions are also required for effectively implementing the emissions trading system. Presently, institutions such as CPCB and the different SPCBs have not effectively implemented the environmental policies due to financial and technical constraints. The justification for technology push policy has itself been the relative ease in implementation. However, as seen in the analysis it also results in significant compliance costs. Against this, emissions trading results in 45 per cent cost savings over a 30-year period. Therefore, there is a need to develop new institutions and strengthen existing ones to effectively implement a cost-effective measure to control $SO_2$ emissions.

*Inter-linkages between Energy-Electricity-Emissions Market*

There are challenges to the development of an emissions trading market in India due to distortions in the energy and electricity markets. This implies that adopting measures to improve the economic efficiency of the emissions market need not necessarily be effective due to the inter-linkages with the energy and electricity markets. For instance, there is administered pricing in the electricity markets. This implies that it could be possible for plants to pass on their compliance costs to the consumers, which would negate the purpose of an emissions trading system.

*Inter-linkages between Environment-Technology Markets*

The success of an emissions trading regime depends on the presence of alternate compliance technologies and know-how. This implies the need for a well-developed technology market in the country that would enable the emitters to choose between alternate technologies available in the market. In the absence of such choices, an emissions trading system would not be different from a technology-push policy. As mentioned earlier, an emissions trading system would provide impetus to further development of technology markets; but specific policies that promote the growth of this market are also necessary.

## CONCLUSION

Electricity reforms in India have benefitted the environment, but indirectly. This is reflected in the case studies of the states of Andhra Pradesh and Gujarat where there has been some decline in $CO_2$

and $SO_2$ emissions after the introduction of the electricity reforms. However, this decline is due to changes in fuel mix, improved technology, and improvement in energy efficiency (reflected in the improved heat rates) in the power plants of these states following the reforms. Environmental concerns have not been specifically addressed in the reforms process. Thus, there is a case for alignment of the electricity and the environmental markets. Another insight obtained from these studies is that newer plants are more environmentally friendly than older plants, which is reflected in their marginal emission intensity curves that are steeper than the overall curves. Thus, indirect factors in the reform process have contributed to environmental improvement, more so for local pollutants. Besides, there are also state-wise variations in emissions of both global and local pollutants, as reflected in the differences in the overall emission baselines of the two states. A primary reason for this difference is the difference in fuel mix of these two states (AP has more hydro, but the reduced generation from hydro raises the baselines).

Presently, national-level policies for controlling pollutant emissions from power plants rely on command-and-control instruments which mandate the use of specific technologies. However, such technology-push policies do not provide an optimal solution, since they do not take into account the differences in marginal abatement costs between different power plants. Therefore, it is necessary for policy-makers to consider alternate policy tools, such as emissions trading, which is more cost effective. When compared to a technology-push policy, an emissions trading regime generates an annual average cost-savings of Rs. 3.6 billion for equivalent reductions. These savings should be invested in clean-coal technologies, strengthening institutions for effective policy implementation and other air quality management measures. Such an approach would align the economic considerations of power plants with environmental interests. However, it is necessary to precisely work out the details regarding the actual implementation of this regime (in terms of policies, implementing agencies, regulators and so on).

The energy, electricity and environment sectors have strong linkages. The consumption of energy resources generates negative externalities in the form of environmental degradation, including air pollution. Power plants are major contributors to local and global pollutant emissions in India, mainly because of the dominance of coal as a fuel in these plants. Policy-makers are more concerned with managing the local pollutants due to their damaging impacts on human health. Given the strong linkage of electricity sector with the environment, it is necessary to align environmental policies in the electricity reform process. The same is true for the energy sector, since energy costs is nearly 80-90 per cent of electricity costs. Thus, aligning the energy-electricity and environment markets is crucial for the success of the reforms in the electricity sector. An important step in the direction of energy and electricity market alignment is regional cooperation among the South-East Asian countries (Nair, et al., 2003). In the case of India, such a cooperation will improve the energy markets, for instance the availability of gas will improve, which in turn will affect the electricity market by providing more fuel choice in generation.

The complex interplay between different sectors and related institutions implies that policies in a specific sector would not achieve the desired impacts without complementary policies in related sectors. Sequencing and pacing of reforms is also crucial, given the resource constraints of the government (Stiglitz, 1999). Thus, an integrated approach that takes into consideration the multi-sectoral linkages should be adopted in electricity reforms.

## NOTES

1. Heat rate is defined as the energy in kilo calories (kcal) required to produce a kilowatt-hour (KWh) of electricity. Lesser heat rate signifies lower energy requirement to produce the same unit of electricity and hence better efficiency.
2. The putty-clay model defines capital to be putty before investment but clay once the investment has been made.
3. Units for plants other than gas and naphtha have been defined as one boiler turbine generator (BTG) group. For gas and naphtha plants, due to the presence of multiple gas units and single steam unit, a thermally interconnected system of gas and steam units have been defined as a group.
4. In IGCC technology, 99 per cent of the $SO_2$ removal takes place insitu (WB, 1997).

## REFERENCES

Alagh, Y.K., J. Shah and V.K. Shah, *Power Economics in Gujarat*, New Delhi: Har-Anand Publications, 1998.

AP GENCO (Andhra Pradesh Power Generation Corporation Limited), *Generation Particulars*, www.apgenco.com, 30 August 2003.

AP TRANSCO (Transmission Corporation of Andhra Pradesh Limited), *Power Development in Andhra Pradesh 2000-01*, 2001.

Boubakri, N. and J.C. Cosset, 'The Financial and Operating Performance of Newly Privatized Firms: Evidence from Developing Countries', *Journal of Finance*, 53, June 1998.

CMIE (Center for Monitoring Indian Economy), *Energy*, Mumbai: CMIE, 1999.

———, *Energy*, 2001.

———, *Energy*, 2002a.

———, *Prowess: The Indian Corporate Database*, 2002b.

———, *Energy*, 2003a.

———, *Monthly Review of the Gujarat Economy*, 2003b.

———, *National Income Statistics*, 2004a.

———, *Monthly Review of the Andhra Pradesh Economy*, 2004b.

Chandler, W., R. Schaeffer, Z. Dadi, P.R. Shukla, F. Tudela, O. Davidson and S. Alpan-Atamer, *Climate Change Mitigation in Developing Countries: Brazil, China, India, Mexico, South Africa, and Turkey*, Arlington, Virginia, US, Pew Center on Global Climate Change, 2002.

CPCB (Central Pollution Control Board), *Environmental Management in Selected Industrial Sectors: Status and Needs*, New Delhi: Ministry of Environment and Forests, Government of India, 2003.

Garg, A., P.R. Shukla, S. Bhattacharya and V.K. Dadhwal, 'Sub-region (district) and Sector Level $SO_2$ and $NO_X$ Emissions for India: Assessment of Inventories and Mitigation Flexibility', *Atmospheric Environment* (35), 2001a, pp. 703-13.

_____, S. Bhattacharya, P.R. Shukla and V.K. Dadhwal, 'Regional and Sectoral Assessment of Greenhouse Gas Emissions in India', *Atmospheric Environment* (35), 2001b, pp. 2679-95.

_____, M. Kapshe, P.R. Shukla and D. Ghosh, 'Large Point Source (LPS) Emissions from India: Regional and Sectoral Analysis', *Atmospheric Environment* (36), 2002a, pp. 213-24.

_____, and P.R. Shukla, *Emissions Inventory of India*, New Delhi: Tata McGraw-Hill, 2002b, p. 278.

_____, et al., 'Future GHG and Local Emissions for India: Policy Links and Disjoints', *Mitigation and Adaptation Strategies for Global Change*, vol. 8, no. 1, 2003, pp. 71-92.

_____, (Government of India), *Economic Survey of India 1990-1991*, New Delhi: Ministry of Finance, Government of India, 1991.

_____, *Economic Survey of India 1991-1992*, New Delhi: Ministry of Finance, Government of India, 1992.

_____, *Economic Survey of India 1992-1993*, New Delhi: Ministry of Finance, Government of India, 1993.

_____, *Economic Survey of India 1993-1994*, New Delhi: Ministry of Finance, Government of India, 1994.

_____, *Economic Survey of India 1994-1995*, New Delhi: Ministry of Finance, Government of India, 1995.

_____, *Economic Survey of India 1995-1996*, New Delhi: Ministry of Finance, Government of India, 1996.

_____, *Economic Survey of India 1996-1997*, New Delhi: Ministry of Finance, Government of India, 1997.

_____, *Economic Survey of India 1997-1998*, New Delhi: Ministry of Finance, Government of India, 1998.

_____, *Economic Survey of India 1998-1999*, New Delhi: Ministry of Finance, Government of India, 1999.

_____, *Economic Survey of India 1999-2000*, New Delhi: Ministry of Finance, Government of India, 2000.

_____, *Economic Survey of India 2000-2001*, New Delhi: Ministry of Finance, Government of India, 2001.

_____, *Economic Survey of India 2001-2002*, New Delhi: Ministry of Finance, Government of India, 2002.

Heracleous, L., 'State Ownership, Privatization and Performance in Singapore: An Exploratory Study from a Strategic Management Perspective', *Asia Pacific Journal of Management*, 18, 2001, pp. 69-81.

IPCC (Intergovernmental Panel on Climate Change), 'Guidelines for National Greenhouse Gas Inventories: Workbook', 1996.

Kapshe, M., Amit Garg and P.R. Shukla, 'Application of AIM/Local Model to India using Area and Large Point Sources', in M. Kainuma, Y. Matsuoka and T. Morita (eds.), *Climate Policy Assessment: Asia-Pacific Integrated Modelling*, Tokyo: Springer-Verlag, 2002.

Kapshe, Manmohan, 'Integrated Assessment of Energy and Environment Future: Linking Policy to Implementation', Thesis for Fellow Programme in Management (unpublished), Ahmedabad: Indian Institute of Management, 2003.

Khanna, M. and D. Zilberman, 'Barriers to Energy Efficiency in Electricity Generation in India', *The Energy Journal*, vol. 20, no. 1, 1999, pp. 25-41.

———, 'Adoption of Energy Efficient Technologies and Carbon Abatement: The Electricity Generation Sector in India', *Energy Economics*, 23 (2001), pp. 637-58.

Lovei, M. and B.S. Gentry, 'The Environmental Implications of Privatization', The World Bank, Paper No. 426, 2002.

Ministry of Coal, http://coal.nic.in, November 2003.

Ministry of Petroleum and Natural Gas, http://petroleum.nic.in, November 2003.

Ministry of Power, 'Blue Print for Power Sector Development, Government of India', August 2001.

Nellis, J., 'Is Privatization Necessary?', *Viewpoint*, no. 7, World Bank, Washington D.C., 1994.

Pandey, Rahul and P.R. Shukla, 'Methodology for Exploring Co-benefits of $CO_2$ and $SO_2$ Mitigation Policies in India using AIM/Enduse Model', in M. Kainuma, Y. Matsuoka, and T. Morita (eds.), *Climate Policy Assessment: Asia-Pacific Integrated Modeling*, Springer-Verlag, 2003.

Pollitt, M.G., *Ownership and Performance in Electric Utilities*, Oxford: Oxford Institute of Energy Studies/Oxford University Press, 1995.

Nair, R., P.R. Shukla, M. Kapshe, A. Garg and A. Rana, 'Analysis of Long-term Energy and Carbon Emission Scenarios for India', *Mitigation and Adaptation Strategies for Global Change*, Kluwer Academic Publishers, vol. 8, no. 1, pp. 53-69.

Shukla, P.R., 'Biomass Energy in India: Transition from Traditional to Modern', *The Social Engineer*, vol. 6, no. 2, 1997.

Shukla, P.R., S.K. Sharma and P. Venkata Ramana, *Climate Change and India*, New Delhi: Tata McGraw-Hill, 2002, p. 317.

Shukla, P.R. et al., 'Impact of Power Sector Reforms on Technology, Efficiency and Emissions: Case Study of Andhra Pradesh, India', Working Paper no. 20, PESD, Stanford University, 2004a.

_____, 'Impact of Power Sector Reforms on Technology, Efficiency and Emissions: Case Study of Gujarat, India', Working Paper no. 21, PESD, Stanford University, 2004b.

_____, A. Rana, A. Garg, M. Kapshe and R. Nair, *Climate Policy Assessment for India: Applications of Asia-Pacific Integrated Model (AIM)*, Hyderabad: Universities Press, 2004c.

Singh, J., 'Plant Size and Technical Efficiency in the Indian Thermal Power Industry', *Indian Economic Review*, vol. XXVI, no. 2, 1991, pp. 239-52.

Stiglitz Joseph, E., 'Whither Reform? Ten Years of Transition', paper prepared for the Annual Bank Conference on Development Economics, Washington D.C., 28-30 April 1999.

UNEP, UCCEE and UNCTAD, *An Emerging Market for the Environment: A Guide to Emissions Trading*, Denmark: United National Publications, 2002.

Verhoef, E.T. and P. Nijkamp, 'Second-best Energy Policies for Heterogeneous Firms', *Energy Economics*, 21, 1999, pp. 111-34.

Zhang, C., Michael M. May and Thomas C. Heller, 'Impact on Global Warming of Development and Structural Changes in the Electricity Sector of Guangdong Province, China', *Energy Policy*, 29, 2001, pp. 179-203.

# 14

## *The Act as a Base for Rethinking Technology and Managerial Mixes*

JOËL RUET

The long neglect of maintenance, quality of service, cost-efficiency and conservation is the result of an over-centralized, over-technical administration, along with an absence of technical innovation, paradigm. Shifting from one paradigm to another always requires a catalytic element, as well as bridging the political lock-ins. We address here two dimensions: on the one hand, the Act may serve a catalytic role if its drawbacks are overcome, while new technologies or better mix of technologies can ease the transition process.

A legislation like the Electricity Act provides transfer of critical decision-making functions to independent agencies. Thus, it separates policy making and regulations. Many of the solutions call for investments for augmenting systems capacity, technological upgradation, modern operating procedures, efficiency improvements, new managerial culture in SEBs or their successor entities and responsible customer behaviour (Subba Rao). The crux of the reforms thus lies in the capacity to create a momentum among various stakeholders. Here, the history of civil society involvement in public governance is a new one in India. Neither some risks of excluding the poor, nor the necessity to explore technical issues and energy choices are absent from the Act and its institutional environment (Mahalingam, Ruet, Menon-Chaudhuri et al.). As Kalra et al. recall, benchmarking of the electricity distribution companies in India is still a weak point in the present scenario, and planning of capital investment, load growth, better quality of supply all relate to performance of the sector.

Technically, both the means to generate and distribute energy, and the ways to use it have diversified. An electric system can be configured in many ways according to the use by its clients, while electric devices cover today a wide range and, to say the least, a wide dispersion in quality. Electricity is no longer a mono-dimensional good and service. It implies a variety of specific relations, and thus implies controlling both the demand and the supply.

This implies energy saving and energy efficient equipments have to be promoted, with most likely a role in this for the regularity commissions (through tariff incentives and regulation of device quality). This encompasses use of efficient lights, fans, stabilizers. . . . Similarly, suitable voltage compensators and better continuity of supply may render void the need for voltage stabilizers and UPS. Along with the huge potential of savings through grid optimization and reactive capacity addition, these short-term measures are the best savings potential for India's energy sector and the best cushion for higher social use of energy.

The proper enforcement and monitoring of these measures will of course call for a better information-based decision-making system. Congestion management and operation of the network through better use of integrated resource planning are ways to urgently explore, as these involve minimum investment. As a case in point, India can use its Information Technology companies to help develop such an innovative scenario.

These short-term measures, also relieve pressure on the longer-term energy choices. No doubt all scenarios must be explored, but time is still needed to find ways to implement proper and more demanding security needs, as well as ensure pollution control and social rehabilitation for nuclear, coal and hydro power. These are problems that every country has to address, the hydro power however being more stringent for densely populated India.

The future is set, at least if one uses the best techno-economic mix available today. But where do we stand in terms of implementation? At the end of 13-odd years' journey of reforms, the Indian power sector is still limping, manifest in three dimensions as far as technology is concerned.

Data envelop analysis shows that the performances of utilities were independent of restructuring strategy, electricity reforms like unbundling, presence of an independent regulatory commission and such other sector-level reforms-related factors. Balasubramanian concludes that the factors that could push efficiencies in the electricity sector could be more fundamental ones like strengthening of institutional set-ups at the state level, and better policy making, planning and governance by the states. He calls for addressing of the diverse needs state-level utilities have to improve performance. He also states that rather than limiting reforms to focus on financial turnaround of the sector, we should adopt a broader perspective by emphasizing on institutional strengthening and better governance. In other words, the base has been established, the necessary phase of preparing the ground, out of which state-specific and situation-specific measures have to be developed.

The second striking point is that, still, at an all-India level, huge untapped margins remain in optimal generation/network and investment/operation mix (Ruet). More than a decade of reforms in this field has not allowed establishing the mechanisms, information and procedures that would allow tapping resources that actually lie at the field level, some of them at the 'last-mile' level of the network, where no centralized planning of the SEBs can ever outbeat or even match a sound decentralization.

On the environmental front now, Menon et al. show that coal-based power plants are big polluters in India, not only for greenhouse gases but also for more local pollutants like $SO_2$. Despite not having a holistic look at this, one can conclude that electricity reforms in India have benefited the environment, though indirectly. This is reflected in the case studies of the states of Andhra Pradesh and Gujarat where there has been some decline in $CO_2$ and $SO_2$ emissions after the introduction of electricity reforms, mostly due to changes in fuel mix, improved technology, and improvement in energy efficiency (reflected in the improved heat rates). Environmental concerns have not been specifically addressed in the reforms process, however. Thus, there is still a case, that will become more pressing with time, of aligning the electricity and the environmental markets.

To put in different words, an institutional reform like the Electricity Act, in India, happens in and has to cater to an environment where technological choices, mixes, implementation modes, energy choices and environmental aspects have remained in an inertial mode for decades, that too in a planned, centralized model. The Act, cannot address all these, as it has to take into account political and legal dimensions. Though it calls for many regulatory as well as managerial and technological developments, the agenda has just been set, not achieved.